SANITARY ENGINEERING PROBLEMS and CALCULATIONS for the PROFESSIONAL ENGINEER

SANITARY ENGINEERING PROBLEMS and CALCULATIONS for the PROFESSIONAL ENGINEER

by

HARRY S. HARBOLD

Environmental Engineer
U.S. Environmental Protection Agency
Philadelphia, Pennsylvania

P.O. Box 1425, 230 Collingwood, Ann Arbor, Michigan 48106

Library of Congress Catalog Card Number 79-88898
ISBN 0-250-40319-6

Manufactured in the United States of America

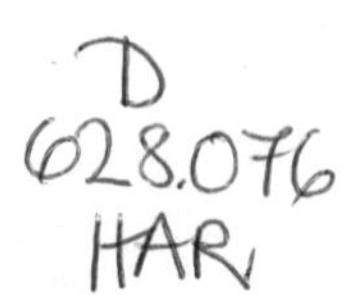

PREFACE

Many textbooks and publications have been written on sanitary and environmental engineering. These texts often have two major drawbacks. Either the publication is too theoretical and does not include practical design applications, or the material covers too narrow a subject area.

This book was written to cover the whole spectrum of sanitary engineering, to present only engineering calculations and omit detailed theoretical discussion. Over one hundred practical engineering problems with detailed solutions are included. The types of problems that are solved simply by using empirical equations, the approach most often included in engineering texts, have purposely been omitted. The reader will notice, for example, that the solution to many problems is based on the analysis and interpretation of engineering data. This approach should make the book especially useful to the practicing engineer. The book is also organized as a study guide for the professional engineer examination, and can be used as a reference book and supplement for college texts.

Four main subject areas are included: fluid flow, water supply and treatment, wastewater treatment, and economics. Chapter 1 emphasizes the application of the Manning, Hazen-Williams, and Darcy-Weisbach equations for water and wastewater conveyance systems. Chapter 2 includes solutions to specific problems associated with water supply and treatment. Chapters 3, 4, and 5 cover the more important wastewater treatment areas relating to biological, physical-chemical, and advanced wastewater treatment technology; sludge handling and disposal; and statistical analysis of data. Chapter 6 provides several different types of economic analyses applicable to sanitary engineering.

In solving these problems certain assumptions are made. The reader should realize that final design of any engineering system should consider site specific conditions, the engineer's own judgment, and equipment manufacturer recommendations. Any local, State, or Federal environmental regulations should also be considered.

Harry Harbold

Harry S. Harbold is an Environmental Engineer with the U.S. EPA, Philadelphia. A Registered Professional Engineer, he received his MS in Environmental Engineering from Drexel University, and his BS in Chemical Engineering from Auburn University.

His responsibilities include engineering plans and specifications review for treatment plant design and construction, and inspection of waste treatment facilities. He is also involved in issuance and enforcement of industrial and municipal NPDES permits based upon existing and future available treatment technology.

Formerly a chemical engineer with the Naval Air Development Center, Warminster, Pennsylvania, his work there entailed laboratory and field investigations of environmental systems, and the determination of toxic effects from corrosion inhibitors used in aircraft wash systems. He also conducted laboratory tests involving reverse osmosis and chemical precipitation to remove toxic chemicals. He has written articles for various professional journals.

CONTENTS

APPENDIXES

Chapter 1

FLUID FLOW

PROBLEM 1.1

Water flows by gravity at a rate of 500 gallons per minute through 5 miles of cast iron pipe from one reservoir to another reservoir located 200 feet below. Calculate the minimum pipe diameter required and the standard size pipe that should be used.

Solution

Assume pipe is flowing full and steady uniform flow. Problem will be solved using three different methods to illustrate use of the Manning formula, Hazen-William formula, and Darcy-Weisbach equation. The Manning and Hazen-Williams equations are limited to turbulent flow and common ambient temperatures for water. The Manning equation is commonly used in sewer design and open channel flow, the Hazen-Williams equation for water supply systems. The Darcy-Weisbach equation is often used for fluids other than water.

Method 1 - Manning formula may be written as

$$V = \frac{1.486}{n} R^{2/3} S^{1/2}$$

where V = velocity, feet/second

n = Manning coefficient

R = hydraulic radius, feet

S = slope of hydraulic grade line, feet/feet

Values of n for the Manning equation are shown in Table 1.1.[1] Values indicated are for good to fair construction. For poor

Table 1.1

Manning Coefficient

Type of Channel	Manning's n Range
Closed Conduit	
Concrete	0.011 - 0.013
Corrugated Metal	0.024
Vitrified Clay	0.012 - 0.014
Cast-iron	0.013
Steel	0.009 - 0.011
Brick	0.014 - 0.017
Open Channels; Lined	
Concrete	
Trowel Finish	0.012 - 0.014
Float Finish	0.013 - 0.015
No Finish	0.013 - 0.017
Brick	0.014 - 0.017
Asphalt	0.013 - 0.016
Gravel Bottom, sides of	
Formed Concrete	0.017 - 0.020
Random Stone	0.020 - 0.023
Riprap	0.023 - 0.033
Open Channels; Excavated	
Earth, uniform	
Clean	0.016 - 0.020
Short Grass	0.022 - 0.027
Dense Weeds	0.030 - 0.035
Channels not maintained	
Weeds and brush	0.080 - 0.120
Street and Expressway Gutters	
Concrete, troweled	0.012
Asphalt Pavement	
Smooth	0.013
Rough	0.016

quality construction, use larger value of n.

Use n = 0.013.

For pipe flowing full hydraulic radius = diameter/4.

Flow rate = 500 gpm = 0.72 mgd = 1.11 cfs

$$\text{Slope} = \frac{200 \text{ ft}}{5 \text{ miles x } 5280 \text{ ft/mile}} = 0.007576 \text{ ft/ft}$$

Write Manning equation as

$$V = \frac{Q}{A} = \frac{1.486}{n} (d/4)^{2/3} S^{1/2}$$

where Q = flow rate, cfs

A = pipe area, ft^2

$$\frac{1.11 \text{ cfs}}{\pi d^2/4} = \frac{1.486}{0.013} (d/4)^{2/3} (0.007576)^{1/2}$$

d = 0.68 feet

d = 8.16 inches minimum pipe diameter

Specify 10-inch nominal diameter cast iron pipe

Method 2 - Hazen-Williams equation may be written as

$$V = 1.318 \text{ C } R^{0.63} S^{0.54}$$

$$\text{or } Q = 0.285 \text{ C } d^{2.63} S^{0.54}$$

where V = velocity, feet/second

Q = flow rate, gpm

C = roughness coefficient

R = hydraulic radius, feet

S = slope of hydraulic grade line, feet/feet

d = pipe diameter, inches

Values of C for Hazen-Williams equation used for design are given in Table 1.2.[2] Use C = 100 and solve for pipe diameter, d.

$$500 \text{ gpm} = 0.285\ (100)\ (d)^{2.63}\ (0.007576)^{0.54}$$

$$d = 8.10 \text{ inches minimum pipe diameter}$$

Specify 10-inch nominal diameter pipe.

Table 1.2

Hazen-Williams Coefficient

Type of Pipe	Value of C New Pipe	 Design
Cement-Asbestos	150	140
Cast iron	130	100
Concrete	120	100
Vitrified	110	100
Corrugated Steel	60	60
Copper, brass, lead, tin or glass pipe and tubing	140	130
Cement-lined iron or steel centrifugally applied	150	140
Wrought-iron	130	100
Wood-stave	120	110

Method 3 - Darcy-Weisbach equation may be written as

$$h_f = f \frac{L}{D} \frac{V^2}{2g}$$

where h_f = friction loss, feet

f = friction factor

L = pipe length, feet

V = velocity, feet/second

g = acceleration due to gravity, feet/sec^2

The value of f varies with the pipe roughness, Reynolds number, and other factors. The Moody diagram[3] is used to determine the value of f. For new cast iron pipe the absolute roughness, ε, equals 0.00085 feet.

Assuming a 10 inch diameter pipe calculate ε/D

$$\varepsilon/D = \frac{0.00085\ ft}{10/12\ ft} = 0.00102$$

Reynolds number can be calculated from the equation,[4]

$$R_e = \frac{50.6\ Q\rho}{d\mu}$$

where R_e = Reynolds number

Q = flow rate, gpm

ρ = density, lbs/ft^3

d = pipe diameter, inches

μ = absolute viscosity, centipoise

For water at 60° F

μ = 1.10 centipoise

ρ = 62.371 lb/ft^3

$$R_e = \frac{50.6 \times 500 \times 62.371}{8 \times 1.10} = 143{,}453$$

From Moody diagram read f = 0.022 for $R_e = 1.4 \times 10^5$ and $\varepsilon/D = 0.00102$.

$$h_f = f\frac{L}{D}\frac{V^2}{2g} = f\frac{L}{D}\frac{(Q/A)^2}{2g} = \frac{8fLQ^2}{\pi^2 d^5 g}$$

$$d^5 = \frac{8 \times 0.022 \times (5 \times 5280) \times (1.11)^2}{32.2 \times (3.14)^2 \times 200}$$

d = 7.45 inches minimum pipe diameter

Since the value of ε used is for new clean pipe, a 10 inch nominal pipe size is recommended to account for increased friction factor after period of use.

PROBLEM 1.2

A tank is to be filled with 50,000 pounds of a slurry (density = 83 lb/cu. ft., viscosity = 25 centipoise) by pumping below the elevation of the discharge pipe. A 3-inch pipe 30 feet long is used from the larger tank to the pump. Fittings include an open globe valve. A 3-inch pipe 100 feet long is used from the pump to the smaller tank. Fittings include a standard tee. A 3 horsepower pump with an overall efficiency of 70% is used. How long will it take to fill the tank?

Solution

Use Darcy-Weisbach equation: $h_f = f \frac{L}{D} \frac{V^2}{2g}$

Value of f varies with Reynolds number, R_e

For laminar flow (R_e <2000) $f = 64/R_e$

For turbulent flow (R_e >4000) $f = \varepsilon/D$

Solve by trial and error. Assume value of flow, Q, and calculate Reynolds number from the equation (Problem 1.1),

$$R_e = \frac{50.6\ Q\rho}{d\mu}$$

Assume Q = 150 gpm

$$R_e = \frac{50.6 \times 150 \text{ gpm} \times 83 \text{ lb/ft}^3}{3 \text{ inches} \times 25 \text{ cp}}$$

$$R_e = 8{,}400$$

Assume piping is commercial steel and use $\varepsilon = 0.00015$

From Moody diagram, $\varepsilon/D = 0.006$, $f = 0.0325$

$$V = \frac{Q}{A} = \frac{150 \text{ gpm} \times 1/60 \times 0.1337 \text{ ft}^3\text{/gal}}{\frac{\pi(3/12)^2}{4}}$$

V = 6.82 ft/sec

From Table 1.3,[5] equivalent length of an open gate valve for 3-inch pipe is 1.7 feet. Equivalent length of standard tee is 5.1 feet. Entrance and exit losses are assumed equivalent to 1.5 $V^2/2g$.

Entrance and exit losses = $2[1.5(6.82)^2/64.4]$

= 2.94

L = 100 + 30 + 1.7 + 5.1 + 2.94

L = 139 feet

For 3-inch pipe,

$$h_f = 0.0325 \times \frac{139 \text{ ft}}{3/12 \text{ ft}} \times \frac{(6.82 \text{ ft/sec})^2}{64.4 \text{ ft/sec}^2}$$

h_f = 13.1 feet

Total head = 13.1 + 17.0 = 30.1 feet

Assume brake horsepower is given in problem.
Solve for Q using equation,

$$HP = \frac{Q \times \text{Head} \times \text{Specific Gravity}}{\text{Efficiency} \times 3960}$$

$$Q = \frac{0.70 \times 3960 \times 3.0 \text{ Hp}}{(83/62.4) \times 30.1 \text{ ft}} = 207 \text{ gpm}$$

Assumed value of Q too low. Try Q = 175 gpm.

Calculate R_e = 9800

ε/D = 0.0006

f = 0.0315

V = 7.95 ft/sec

L = 140 feet

h_f = 17.3 feet

Total head = 34.3 feet

$$Q = \frac{0.7 \times 3960 \times 3.0}{(93/62.4 \times 34.3} = 182 \text{ gpm} \quad \text{(close enough)}$$

$$\text{Pumping time} = 50{,}000 \text{ lbs} \times \frac{\text{ft}^3}{83 \text{ lbs}} \times \frac{7.48 \text{ gal}}{\text{ft}^3} \times \frac{\text{min}}{175 \text{ gal}}$$

= 26 minutes

TABLE 1.3

EQUIVALENT LENGTH OF PIPE FOR VARIOUS FITTINGS

	For Nominal Pipe Diameter Shown, Equivalent Length of Pipe, Ft.																		
	½ In.	1 In.	1½ In.	2 In.	3 In.	4 In.	6 In.	8 In.	10 In.	12 In.	14 In.	16 In.	18 In.	20 In.	24 In.	30 In.	36 In.	42 In.	48 In.
Gate valve, open	0.3	0.6	0.9	1.2	1.7	2.3	3.5	4.5	5.8	6.9	8.0	9.0	10	12	14	17	20	23	27
Gate valve, ½ open	11	17	25	34	50	67	100	135	170	195	230	260	300	330	400	500	600	710	790
Globe valve, open	16	27	43	54	80	110	160	210	280	330	380	430	480	540	670	830	1,000	1,200	1,300
Angle valve, open	9.0	15	22	28	42	57	85	110	140	160	190	220	250	280	340	420	500	600	690
Check valve	3.9	6.5	11	14	19	25	40	51	66	77	90	110	120	130	160	200	230	280	320
Standard tee,	3.4	5.7	9	12	17	22	34	44	57	67	75	89	100	110	130	170	200	240	270
Standard tee,	3.4	5.7	9	12	17	22	34	44	57	67	75	89	100	110	130	170	200	240	270
Standard tee,	1.1	1.7	2.7	3.5	5.1	6.9	11	14	18	20	23	27	30	34	42	52	61	72	82
90° Elbow, std.	1.5	2.7	4.4	5.4	8.1	11	17	21	26	31	37	42	48	52	63	80	98	110	130
90° Elbow, long radius	1.1	1.7	2.7	3.5	5.1	6.9	11	14	18	20	23	27	30	34	42	52	61	72	82
180° Return bend	3.6	6.1	10	13	18	23	36	48	60	72	83	100	110	125	150	180	220	260	300
45° Elbow	0.8	1.2	1.9	2.5	3.7	5.0	7.4	10	13	15	17	19	21	23	30	37	44	52	60
Ordinary entrance,	0.9	1.5	2.3	3.0	4.5	5.9	9.0	12	15	18	20	23	25	30	35	45	52	61	70
Sudden enlargement																			
$d/D = 1/4$	1.5	2.7	4.4	5.4	8.1	11	17	21	26	31	37	42	48	52	63	80	98	110	130
$d/D = 1/2$	1.0	1.5	2.6	3.2	4.8	6.3	10	13	16	19	22	24	28	31	38	47	58	68	78
$d/D = 3/4$	0.3	0.6	0.9	1.2	1.7	2.3	3.5	4.5	5.8	6.9	8.0	9.0	10	12	14	17	20	23	27
Sudden contraction																			
$d/D = 1/4$	0.8	1.2	1.9	2.5	3.7	5.0	7.4	10	13	15	17	19	21	23	30	37	44	52	60
$d/D = 1/2$	0.6	0.9	1.5	1.9	2.8	3.7	5.7	7.5	9.7	12	13	15	17	18	22	27	33	40	45
$d/D = 3/4$	0.3	0.6	0.9	1.2	1.7	2.3	3.5	4.5	5.8	6.9	8.0	9.0	10	12	14	17	20	23	27

PROBLEM 1.3

You have been put in charge of the preliminary investigation and the design of a sanitary sewer system to serve a small town. Briefly outline the steps in completing this job.

Solution

Make recommendations regarding service agreements with other areas, availability of Federal and State financial assistance, required sewer rate structure, and bonding capability to finance project.

Review available topographic maps, aerial photos, and City and State highway plans. Field surveys may be required. Determine the location and capacity of any existing sewers and storm water drains. Determine location of underground utilities, water and gas mains, unusual obstructions, required rights-of-way. Determine profile levels for all streets and nearby ground contours, such as stream beds and ditches. Basement depths should be known. Make soil borings to determine subsurface soil and water conditions.

Develop a general arrangement plan showing sewer lines and areas tributary to each. Locate manholes.

Calculate population projection for 30 to 50 year design life. Determine maximum and minimum sewage flows. Consider the amount and type of industrial waste contribution and variation in flow. Design to minimize ground water infiltration and any inflow. Calculate required sewer size, depth, and grade to handle anticipated flows from each drainage area. Calculate manhole invert elevations. Avoid locating sewer lines in close proximity to water lines and other underground utilities.

Prepare plan and profile drawings. List computations in tabular form for accuracy and easy reference.

PROBLEM 1.4

Determine the depth of flow in a finished concrete trapezoidal channel which discharges 1200 cubic feet per second at a slope of 0.0009. Bottom width is 30 feet and side slope is in the ratio of 3 horizontal to 1 vertical. Assume n = 0.012.

Solution

Solve by trial and error using Manning equation. Let D = depth of channel.

$$\text{Area} = 30D + 2(\tfrac{1}{2}D \times 3D) = 30D + 3D^2$$

$$\text{Perimeter} = 30 + 2\sqrt{10D^2} = 30 + 6.32D$$

$$\text{Hydraulic Radius} = \frac{\text{Area}}{\text{Wetted Perimeter}} = \frac{30D + 3D^2}{30 + 6.32D}$$

$$\text{Manning Equation:} \quad Q = \text{area} \times \frac{1.486}{n} R^{2/3} S^{1/2}$$

Since R = Area/wetted perimeter, rearrange as

$$\frac{Q \times n}{1.486 \times S^{1/2}} = \frac{A^{5/3}}{P^{2/3}}$$

Substituting given values for Q, n, S.

$$\frac{(30D + 3D^2)^{5/3}}{(30 + 6.32D)^{2/3}} = \frac{1200 \times 0.012}{1.49 \times (0.0009)^{1/2}} = 323$$

Assume D = 5 feet

$$\frac{[(30)(5) + (3)(5)^2]^{5/3}}{[30 + (6.32)(5)]^{2/3}} = 367$$

Assume D = 4.5 feet

$$\frac{[(30)(4.5) + (3)(4.5)^2]^{5/3}}{[30 + (6.32)(4.5)]^{2/3}} = 304$$

By straight line interpolation, D = 4.65 feet

PROBLEM 1.5

Three types of pumps used in wastewater treatment and collection systems are (1) axial pump (2) pneumatic ejector (3) reciprocating displacement pump. For each pump (a) Describe the basic mode of operation (b) List common uses

(c) Identify at least two advantages and disadvantages of each.

Solution

Axial Pump

(a) Axial flow pumps are one of the many types of centrifugal pumps. In centrifugal pumps water enters the suction pipe and is discharged by an impeller in the pump casing through the discharge pipe. Energy is converted from velocity head into pressure head. In the axial flow pump the water flow is parallel to the axis of the pump.

(b) Axial pumps can be used for pumping treatment plant effluent and settled raw wastewater influent.

(c) Advantages: Dependable operation with only routine maintenance required; Relatively quiet operation; Provides even pumping flow rate; Does not require a large amount of space.

Disadvantages: Suitable only at low heads; Can not pump liquid containing trash or other solids that may clog pump.

Pneumatic Ejector

(a) Sewage enters a closed tank until a certain level is reached. Air is then injected under pressure to force the water out the discharge pipe.

(b) Pneumatic ejectors are most commonly used in lift stations, to pump treatment plant screenings, and to pump raw sludge and sludge cake.

(c) Advantages: Requires little maintenance, automatically operated; Not easily clogged by solids; Closed system prevents escape of sewer gases.

Disadvantages: Low efficiency; Large volume air needed; Low volume pumping capability.

Reciprocating Displacement Pump

(a) A piston or plunger is used to draw water into the pump cylinder and to discharge the water through the discharge pipe.

(b) Reciprocating pumps may be used for pumping primary sludge, activated sludge, digested sludge, chemical sludge, and for chemical feed systems.

(c) Advantages: Constant capacity against widely varying head; Self priming; Handles large solids; Large discharge head may be provided.

Disadvantages: Not suitable for operation requiring smooth steady discharge; Can handle only low volumes; Additional maintenance may be required for lubricating valves and pistons.

PROBLEM 1.6

A centrifugal pump is used to deliver water through a 10-inch (inside diameter) iron pipe with an equivalent length of 2950 feet. Static head in the system is 30 feet. The pump characteristics are given in Table 1.4.

(a) What will be the discharge, head, and efficiency at the operating point?

(b) What will be the water horsepower and the brake horsepower required to drive the pump at a flow rate of 2500 gpm?

Solution

(a) Pump characteristic curves are shown in Figure 1.1. The head capacity curve and the efficiency curve are drawn from the data in the problem. The system head curve is the sum of the static head in the system, the friction head loss at a given flow rate, and the velocity head. For example, at a flow rate of 1000 gpm,

Static head = 30 feet

Friction head loss = 1.6 ft/100 ft x 2950 ft = 47.2 feet

Table 1.4

Pump Characteristics for Problem 1.6

Capacity, gpm	Head, feet	Efficiency, %
0	325	0
1000	330	49
1500	325	68
2000	323	80
2500	310	87
3000	285	90
3500	260	89
4000	240	85

Velocity head = 0.26 feet

Total system head = 77.46 feet

The operating point is at the intersection of the system curve and the head capacity curve. From the graph read

Discharge = 3450 gpm

Head = 270 feet

Efficiency = 89%

(b) For water at 68° F, the water horsepower, WHP, delivered by any pump can be calculated from the equation

$$WHP = \frac{QH}{3960}$$

where Q = flow, gpm

H = head, feet

The brake horsepower is the power required to drive the pump and is dependent on pump efficiency. Brake horsepower, BHP, can be calculated from the equation,

$$BHP = \frac{QH}{\text{Efficiency} \times 3960} = \frac{WHP}{\text{Efficiency}}$$

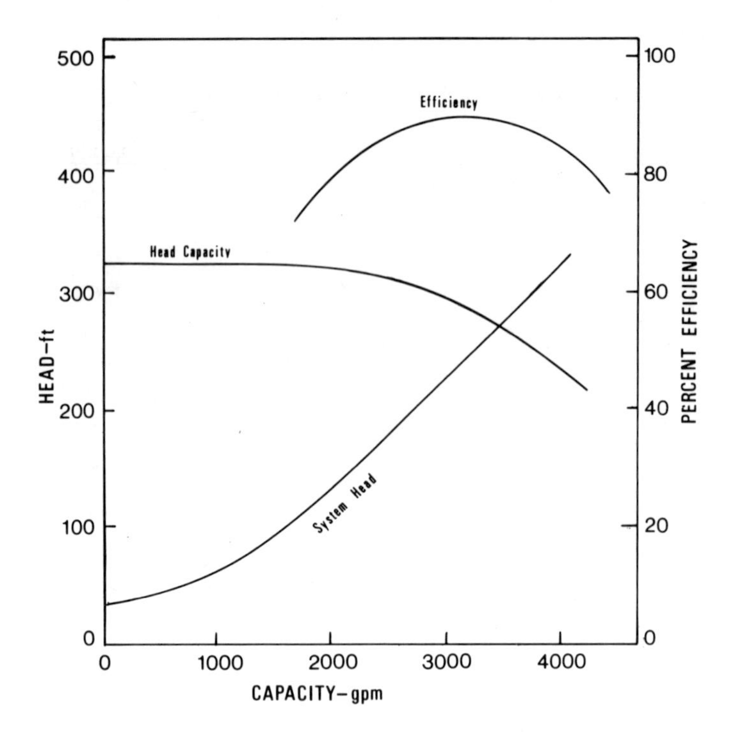

Figure 1.1. Pump curve for Problem 1.6.

From Figure 1.1, at 2500 gpm flow,

Efficiency = 87% Head = 180 feet

$$\text{Water Hp} = \frac{2500 \text{ gpm} \times 180 \text{ ft}}{3960} = 113.6 \text{ Hp}$$

$$\text{Brake Hp} = \frac{113.6 \text{ Hp}}{0.87} = 130.6 \text{ Hp}$$

PROBLEM 1.7

Water flows through a conveyance system consisting of two pipes, A and B, in series. Total head loss through the two pipes is 28 feet. Pipe A is 12 inches in diameter and 1200 feet long. Pipe B is 8 inches in diameter and 600 feet long. It is proposed to add a third pipe, C, to extend the system an additional 900 feet. Determine the diameter of pipe C to maintain the present flow rate if the total head loss through the three pipes can not exceed 40 feet. Assume that f = 0.021, 0.020, and 0.022 for pipes A, B, and C respectively.

Solution

Determine flow rate, Q, through existing pipes A and B. Assume values of Q and solve for head loss through each pipe using Darcy-Weisbach equation. Total head loss equals 28 feet.

Assume Q = 3cfs

Pipe A:

$$h_f = f \frac{L}{D} \frac{(Q/A)^2}{2g}$$

$$h_f = 0.021 \times \frac{1200}{12/12} \times \frac{(3/0.785)^2}{64.4}$$

h_f = 5.72 feet

Pipe B:

$$h_f = 0.020 \times \frac{600}{8/12} \times \frac{(3/0.349)^2}{64.4}$$

h_f = 20.65 feet

Total head loss = 26.37 feet (too low)

Assume Q = 3.1 cfs

Pipe A: h_f = 6.1 feet

Pipe B: h_f = 22.0 feet

Total = 28.1 feet (close enough)

Total head loss through extended system cannot exceed 40 feet. Calculate head loss for pipe C.

Head loss = 40.0 feet - 28.0 feet = 12.0 feet

Calculate pipe diameter using Darcy-Weisbach equation with flow rate of 3.1 cfs.

$$h_f = f \frac{L}{D} \frac{(Q/A)^2}{2g}$$

$$12.0 = 0.022 \times \frac{900}{D} \times \frac{(3.1/A)^2}{64.4}$$

$D = 0.83$ ft = 10 inches

Therefore 10-inch diameter pipe is required.

PROBLEM 1.8

Water flows at a depth of 10 feet in a rectangular canal 13 feet wide excavated in coarse soil. The slope is 8 feet/1000 feet. Determine the diameter of a concrete pipe, flowing full, that will have the same capacity when laid on the same slope.

Solution

Solve using Manning equation.

Assume $n = 0.035$ for earth canal, $n = 0.015$ for concrete pipe.

For Earth Canal,

Area = 10 ft x 13 ft = 130 ft^2

Wetted Perimeter = 13 ft + (2 x 10 ft) = 33 feet

$$\text{Hydraulic Radius} = \frac{\text{Area}}{\text{Wetted Perimeter}}$$

$$= \frac{130\ ft^2}{33\ ft} = 3.94 \text{ feet}$$

Flow Rate, Q, for earth canal equals flow rate for concrete pipe.

Write Manning equation as,

$$Q = \text{area} \times \frac{1.486}{n} = R^{2/3}\ S^{1/2}$$

$$(130)(1.486/0.035)(3.94)^{2/3}\ (0.008)^{1/2} = \frac{\pi d^2}{4}\ (1.486/0.015)\ (d/4)^{2/3}\ (0.008)^{1/2}$$

Concrete pipe diameter = 9.86 feet

PROBLEM 1.9

A 42 inch diameter concrete outfall conduit 500 feet long discharges into an estuary. At high tide, the sewer outlet is submerged a depth of five feet above the invert. If the beginning invert is at 301.6 and the outlet invert is 295.3, determine the capacity and velocity of the sewer outfall at high tide when flowing full. Assume $n \approx 0.015$.

Solution

At high tide calculate hydraulic slope

$$\text{Slope} = \frac{301.6 - (295.3 + 5.0)}{500 \text{ ft}} = 0.0026 \text{ ft/ft}$$

Use Manning Equation: $V = \frac{1.486}{n} R^{2/3} \; S^{1/2}$

$$V = \frac{1.486}{0.015} (0.875)^{2/3} (0.0026)^{1/2} = 4.62 \text{ ft/sec}$$

$$Q = V \times A = 4.62 \text{ ft/sec} \times 9.62 \text{ ft}^2 = 44 \text{ cfs}$$

PROBLEM 1.10

The following data is available for design of a wastewater pumping station:

Force Main discharge elevation 626.52 feet

Pump on elevation 589.85 feet

Pump off elevation 588.50 feet

Center line pump suction 601.2 feet

F.M. discharge length 1650 feet

F.M. discharge fittings (outside station)

1- 45° elbow

1- 90° elbow

Station discharge piping 6" cast iron

Station suction piping 6" cast iron

Flow rate 200 gpm

Total length suction piping 14 feet

Pump elevation 1300 feet

Pump suction line fittings

1- 90° elbow

Pump discharge line fittings

1- gate valve (fully open)

1- 90° elbow

1- check valve

(a) Calculate total dynamic head.

(b) Calculate net positive suction head.

Solution

Use center line of pump as a datum reference. Assume C = 100.

Calculate total dynamic suction lift.

Static suction lift = 601.2 - 588.5 = 12.7 feet

Equivalent length suction piping

Pipe length	=	14 ft
1- 90° elbow	=	17 ft
Total		31 ft

Friction head loss

0.616 ft/100 ft x 31 ft = 0.19 feet

Total = 12.89 feet

Calculate total discharge head.

Static discharge head = 626.52 - 601.2 = 25.32 feet

Equivalent length discharge pipe

Pipe length	=	1650 ft
1- gate valve	=	3.5 ft
1- check valve	=	40.0 ft
2- 90° elbows	=	34.0 ft
1- 45° elbow	=	7.4 ft
Total	=	1734.9 ft

Friction head loss

0.616 ft/100 ft x 1735 ft = 10.69 feet

Total head loss = 36.01 feet

Total dynamic head = total dynamic suction lift and total discharge head

= 48.9 feet

(b) Net positive suction head (NPSH) can be calculated using the formula[6]

$$\text{NPSH, ft} = (H_A - H_{vp}) \pm (H_S - H_F)$$

where H_A = pressure on fluid in suction tank, feet

H_{vp} = vapor pressure of the fluid at the pumping temperature, feet

H_S = static head or lift from the fluid level to the pump datum, feet

H_F = friction head for the design flow rate at the pump suction, feet

At elevation of 600 feet, H_A = 32.98 feet

Vapor pressure, 20° C = 17.535 mm Hg = 0.782 feet water.

H_S = 12.70 feet

H_F = 0.19 feet

$$\text{NPSH} = (32.98 - 0.782) - (12.70 - 0.19) = 19.69 \text{ feet}$$

PROBLEM 1.11

Water flows from a reservoir to a distribution system as shown in Figure 1.2. The reservoir is at a piezometric level of 604.5 feet. The discharge pressure at the end points in the system must be at least 90 psi. The pressure at points 1 and 2 must be at least 110 psi. Calculate the maximum discharge through each pipe. Ground elevation for each point in the system is given as follows:

Point	Elevation, Feet
1	270
2	265
3	250
4	255
5	258
6	245
7	250

Solution

Piezometric level = ground elevation + pressure (feet water)

Hazen-Williams equation may be written in the form[7]

$$f = 0.2083\ (100/C)^{1.85}\ \frac{q^{1.85}}{d^{4.8655}}$$

where f = friction head loss, ft/100 ft pipe

d = inside pipe diameter, inches

q = flow rate, gpm

C = roughness coefficient

Equation may be solved for q knowing slope (difference in piezometric level) and diameter for each pipe. C is assumed equal to 140. Results are tabulated in Table 1.5.

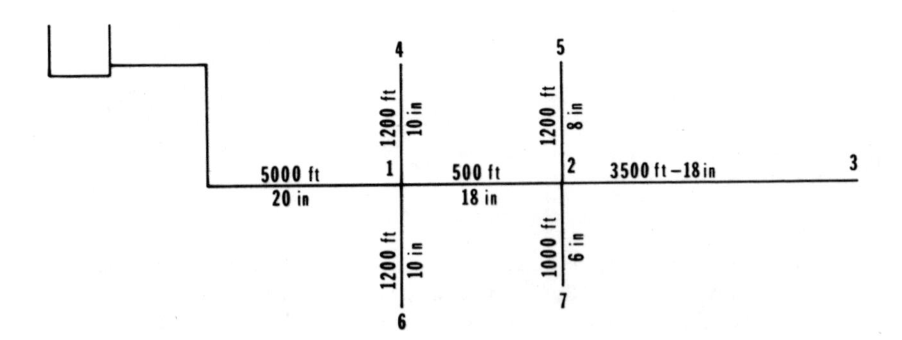

Figure 1.2. Distribution system for Problem 1.11.

Table 1.5

Results for Problem 1.11

Pipe	Head Loss ft/100 ft	Flow, gpm
Reservoir-1	1.63	11,241
1-4	5.08	3,358
1-6	5.92	3,646
1-2	1.00	6,543
2-5	4.42	1,731
2-7	6.10	967
2-3	1.74	8,826

PROBLEM 1.12

A pump draws water through a 10 inch suction line and discharges 2.5 cubic feet per second through a 6 inch pipe. A gauge on the suction line indicates a vacuum of 5 inches of mercury and is located 6 feet below the centerline of the pump. The discharge gauge shows a pressure of 20 psi and is 8 feet below the pump. The power input to the pump is 15 kilowatts. Friction loss between gauges is 3 feet. Determine the efficiency of the pump.

Solution

Solve using Bernoulli equation.

$$P_1 + \frac{V_1^2}{2g} + Z_1 + W_p = P_2 + \frac{V_2^2}{2g} + Z_2 + h_f$$

where P = pressure, feet water

V = velocity, feet/second

g = acceleration due to gravity, ft/sec^2

h_f = friction loss between points 1 and 2, feet

Z = distance from datum, feet

W_p = pump work, feet

Point 1 is defined as the suction line and point 2 the discharge line.

$$P_1 = 5 \text{ inches Hg} \times 1.136 \text{ ft water/inch Hg} = 5.68 \text{ feet}$$

$$P_2 = \frac{20 \text{ lb/in}^2 \times 144 \text{ in}^2/\text{ft}^2}{62.4 \text{ lb/ft}^3} = 46.2 \text{ ft}$$

$$V_1 = Q/A = 2.5 \text{ cfs}/0.545 \text{ ft}^2 = 4.59 \text{ ft/sec}$$

$$V_2 = 2.5 \text{ cfs}/0.196 \text{ ft}^2 = 12.76 \text{ ft/sec}$$

$$-5.68 \text{ ft} - 6.0 \text{ ft} + (4.59)^2/64.4 + W_p = 46.2 \text{ ft} - 8.0 \text{ ft} + (12.76)^2/64.4 + 3.0 \text{ ft}$$

$$W_p = 55.0 \text{ feet}$$

$$\frac{55.0 \text{ ft} \times 2.5 \text{ cfs} \times 62.4 \text{ lb/ft}^3}{\frac{550 \text{ ft-lb/sec}}{\text{Hp}}} = 15.6 \text{ Hp}$$

$$\text{Pump efficiency} = \frac{15.6 \text{ Hp}}{15 \text{ Kw} \times 1.341 \text{ Hp/Kw}} \times 100$$

$$= 77.6\%$$

PROBLEM 1.13

A 12 inch sewer is to discharge 0.5 cubic feet per second at a velocity equivalent to the velocity of the same size sewer flowing full at a slope of 0.3%. Find the velocity and required slope of the 12 inch sewer discharging 0.5 cubic feet per second.

Solution

Solve using Manning equation with n = 0.013 (assumed).

For 12-inch sewer flowing full, slope at 0.3% = 3 feet/1000 feet.

From Manning equation,

$$Q = 1.9 \text{ cfs}$$

$$\text{Velocity} = 2.5 \text{ ft/sec}$$

$$\frac{Q}{Q_{full}} = \frac{0.5}{1.9} = 0.263$$

For sewer flowing partially full as indicated,

$$\text{Velocity} = 0.82 \times 2.5 \text{ ft/sec} = 2.05 \text{ ft/sec}$$

$$\text{Hydraulic Radius, } R = 0.79 \times 0.25 \text{ ft} = 0.1975 \text{ feet}$$

Using Manning equation,

$$\text{Velocity} = \frac{1.486}{n} R^{2/3} S^{1/2}$$

$$2.05 = \frac{1.486}{0.013} (0.1975)^{2/3} S^{1/2}$$

$$S = 0.0028 \text{ ft/ft}$$

PROBLEM 1.14

Using the Marston formula for calculating loads on buried pipes, determine the following.

(a) The load on a 30 inch vitrified clay pipe in a trench under 12 feet of soil backfill. The trench width is five feet.

(b) The minimum required load factor and type of bedding for a safety factor of 1.5.

Solution

(a) Marston's formula may be written as

$$W_c = C_d \, w \, B_d^2$$

where W_c = load on pipe, lb/ft

w = unit weight of backfill soil, lb/ft^3

B_d = width of trench at top of pipe, feet

C_d = dimensionless load coefficient

Since no information is given on the soil density or characteristic, assume a design value of w of 125 lbs/cubic foot.

Determine C_d as 1.7 based on fill depth and trench width.

$$W_c = 1.7 \times 125 \times (5)^2 = 5313 \text{ lb/ft}$$

(b) $$\text{Load factor} = \frac{\text{Design Load}}{\text{Crushing Strength}}$$

Design Load = 5313 lb/ft x 1.5 (safety factor)

= 7970 lb/ft

The crushing strength of 30 inch diameter clay pipe as measured by the three-edge bearing test is 5000 lb/ft.

$$\text{Load factor} = \frac{7970}{5000} = 1.6$$

Class B bedding is recommended.

PROBLEM 1.15

The following equations have been developed to describe a 30 minute unit hydrograph for an urban watershed.[8]

$$T_R = 13.12\ L^{0.315}\ S^{-0.0488}\ I^{-0.490}$$

$$Q = 3.54 \times 10^4\ T_R^{-1.10}\ (A/640)$$

$$T_B = 3.67 \times 10^5\ (A/640)^{1.14}\ Q^{-1.15}$$

$$W_{50} = 4.14 \times 10^4\ (A/640)^{1.03}\ Q^{-1.04}$$

$$W_{75} = 1.34 \times 10^4\ (A/640)^{0.92}\ Q^{-0.94}$$

The following parameters describe the hydrograph.

T_r = time in minutes from the beginning of surface runoff to the peak runoff.

Q = peak discharge, cfs

T_B = base time, the time in minutes from the beginning to the end of surface runoff.

W_{50} = time, in minutes, between the points on the hydrograph when the discharge represented by Q_{50} is half the peak discharge.

W_{75} = time, in minutes, between the points on the hydrograph when the discharge represented by Q_{75} is three-fourths of the peak discharge.

and the following parameters describe the watershed.

A = size of the study area or subarea, acres

L = length of the main drainage channel, feet

S = slope of the main drainage channel, feet/feet

I = percent of impervious cover for the study area or subarea

A one year storm of 30 minutes duration for the study area in question will contribute a total of 0.8 inches of rainfall, or an average rainfall intensity of 1.6 inches per hour. Calculate the size of storm sewer laid on a slope of 0.03 feet/feet necessary to carry the peak storm flow from the drainage area. If the storm sewer is designed to carry only 50 percent of the peak storm flow, with the remainder being diverted to a detention basin, calculate the storage volume required for the basin and sewer diameter. The following information is available to describe the drainage area:

Area = 402 acres

Length of main drainage channel = 7200 feet

Slope of main drainage channel = 0.0354 feet/feet

Percent impervious cover = 25%

Solution

The given values are substituted into the above equations

and the hydrograph characteristics are determined as follows:

T_R = 52 minutes

Q = 288 cfs

T_B = 321 minutes

W_{50} = 71 minutes

W_{75} = 43 minutes

For this case, a unit hydrograph represents the hydrograph of one inch of direct runoff from a 30 minute storm. Therefore, the implied rate of runoff is two inches per hour. The unit hydrograph is modified by multiplying the ordinates of the hydrograph by the ratio:

$$\frac{1.6 \text{ inches/hour}}{2 \text{ inches/hour}} = 0.8$$

The resultant hydrograph is shown in Figure 1.3.

The peak storm flow from the drainage area is equal to:

$$288 \text{ cfs} \times 0.8 = 230 \text{ cfs}$$

Using the Manning equation with n assumed equal to 0.024 and a slope of 0.03 feet/feet,

diameter = 58 inches (flowing full)

Select a 60-inch diameter pipe.

If 50% of the peak storm flow is carried by the storm sewer,

Q = 115 cfs

Required pipe diameter = 29 inches

A 30-inch diameter pipe can be selected.

By graphical integration of the hydrograph in Figure 1.3, the amount of storm water diverted when Q exceeds 115 cfs equals approximately 320,000 cubic feet.

Storage volume = 320,000 ft^3 = 2,400,000 gallons

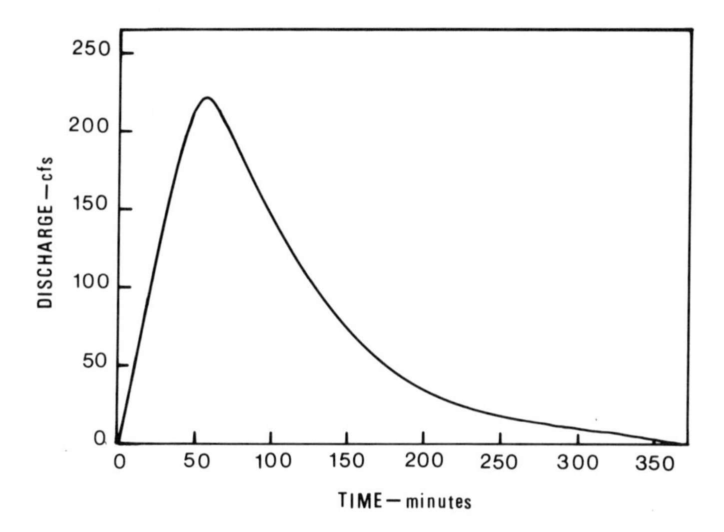

Figure 1.3. Hydrograph for Problem 1.15.

PROBLEM 1.16

An 8-inch sanitary sewer flowing full at a grade of 4 feet per 1000 feet discharges into a steeper 10-inch sewer that must be designed to carry three times the flow of the 8-inch sewer. Determine the invert drop in the transition and the minimum slope for the 10-inch sewer flowing full.

Solution

Solve using Manning equation. Assume n = 0.013.

8-inch Sewer:

Velocity = 2.17 ft/sec

Capacity = 0.76 cfs

Velocity head = $V^2/2g$ = 0.075 feet

10-inch Sewer:

Capacity = 3 x 0.76 = 2.28 cfs

Velocity = 4.18 ft/sec

Velocity head = 0.274 feet

Slope = 0.011 ft/ft

Calculate invert drop using the equation[9]

$$h_i = (H_2 - H_1) + 0.2 \, \Delta \frac{V^2}{2g}$$

where $H = \text{pipe diameter} + \dfrac{V^2}{2g}$

$$H_1 = 0.67 \text{ ft} + 0.075 \text{ ft} = 0.745 \text{ feet}$$

$$H_2 = 0.83 \text{ ft} + 0.274 \text{ ft} = 1.104 \text{ feet}$$

$$h_i = (1.104 - 0.745) + 0.2\,(0.274 - 0.075) = 0.4 \text{ feet}$$

PROBLEM 1.17

In testing a 10-inch diameter water line two adjacent hydrants located 450 feet apart are isolated. Static pressure at the downstream hydrant is measured as 5 psi lower than at the upstream hydrant. At a flow rate of 2750 gpm the upstream and downstream hydrant pressures are determined to be 55 psi and 45 psi, respectively. Determine the Hazen-William coefficient, C, for the line.

Solution

Convert pressure in psi to feet water.

5 psi = 11.54 feet

55 psi = 126.9 feet

45 psi = 103.8 feet

Slope of hydraulic gradient can be calculated as

$$\frac{126.9 - 103.8 + 11.5}{450} = 0.0769 \text{ ft/ft}$$

From Hazen-Williams equation calculate C.

$$C = Q/(0.285\, d^{2.63}\, S^{0.54})$$

$$C = 2750 \text{ gpm}/[0.285\,(10)^{2.63}\,(0.0769)^{0.54}]$$

$$C = 90$$

PROBLEM 1.18

For the system of reservoirs shown in Figure 1.4 calculate the following:

(a) The minimum elevation of reservoir B so that the level can drop 10 feet and water still flow from reservoirs A and B to C.

(b) If reservoir B is at an elevation of 170 feet, calculate the elevation of reservoir C and flow in each pipe.

The flow out of reservoir A is 20,000 gpm.
The elevation of reservoir A is 200 feet.

Solution

Solve using Darcy-Weisbach equation.

(a) For 36-inch pipe

$$h_f = f \frac{L}{D} \frac{V^2}{2g}$$

$$h_f = 0.015 \times \frac{6000 \text{ ft}}{3 \text{ ft}} \times \frac{(6.31 \text{ ft/sec})^2}{64.4}$$

$$h_f = 18.6 \text{ feet}$$

Let H_B = difference in elevation between reservoirs A and B

If $h_f > H_B$, then reservoirs A and B supply reservoir C.

$h_f < H_B$, then reservoir A supplies reservoirs B and C.

Elevation reservoir B must be greater than

(200 ft - 18.6 ft) + 10 ft = 191.4 feet say 192 feet

(b) Elevation Reservoir A = 200 feet

Elevation Reservoir B = 170 feet

Flow will be from reservoir A to reservoirs B and C.

For 24-inch pipe to reservoir B,

h_f = (200 ft - 170 ft) - 18.6 ft = 11.4 feet

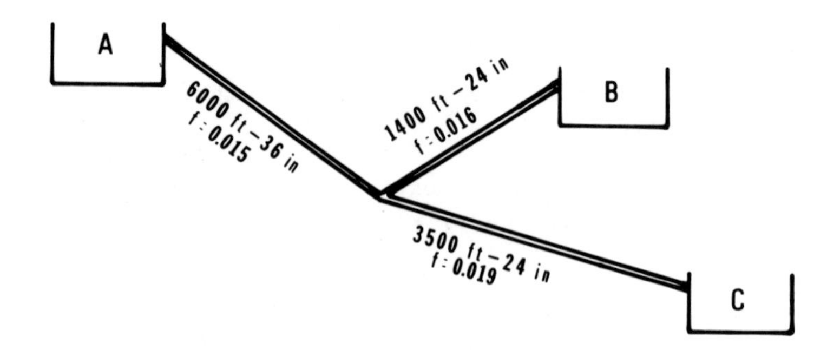

Figure 1.4. Connecting reservoirs for Problem 1.18.

$$11.4 \text{ ft} = 0.016 \times \frac{1400 \text{ ft}}{2 \text{ ft}} \times \frac{(Q/A)^2}{64.4}$$

$$(Q/A)^2 = 65.55$$

$$Q = 25.4 \text{ cfs} = 11{,}400 \text{ gpm}$$

For 24-inch pipe to reservoir C

$$Q = 20{,}000 \text{ gpm} - 11{,}400 \text{ gpm} = 8{,}600 \text{ gpm}$$

Calculate elevation reservoir C. For 24-inch pipe

$$h_f = 0.019 \times \frac{3500 \text{ ft}}{2 \text{ ft}} \times \frac{(6.1 \text{ ft/sec})^2}{64.4} = 19.26 \text{ feet}$$

Elevation reservoir C = 200 ft - (19.26 ft + 18.6 ft)
= 162.1 feet

PROBLEM 1.19

A water supply system supplies a hospital with water at a normal pressure of 40 psi. The supply reservoir may be assumed to be at zero elevation, the storage tank at elevation 100 feet, and the hospital at elevation 50 feet. Discharge pressure of the reservoir is 75 psi. The height of the water in the storage tank is 100 feet above ground elevation. The piping installed between the reservoir and the hospital includes 2000 feet of 10-inch diameter cast iron pipe plus two gate valves, two side outlet tees, and four 45° elbows. The piping between the storage tank and hospital is 4000 feet of 8 -inch cast iron pipe. Assume C = 140 for the pipe.

(a) Calculate the hydraulic head at the storage tank, reservoir, and hospital.

(b) Determine the flow in gallons per minute available for the hospital from the reservoir and storage tank.

(c) Determine the power demand in kilowatts for this operation if the overall efficiency of the reservoir pump and motor is 72%.

Solution

(a) Hospital is located between reservoir and storage tank.

Hydraulic head at reservoir = 75 psi x 2.31 ft/psi

= 173.25 feet

Hydraulic head at hospital = (40 x 2.31) + 50 ft

= 142.4 feet

Hydraulic head at tower = 100 ft + 100 ft

= 200 feet

(b) Calculate flow rates using Hazen-Williams equation with C = 140.

<u>Reservoir to Hospital</u>

Determine equivalent length of pipe.

Pipe length		2000 feet
2 gate valves (½ open)	170 x 2 =	340 feet
2 side outlet tees	57 x 2 =	114 feet
4 45° elbows	15 x 4 =	60 feet
	Total =	2514 feet

$$\text{Head Loss} = \frac{173.25 \text{ feet} - 142.4 \text{ feet}}{2514 \text{ feet}} = 0.0123 \text{ ft/ft}$$

$$Q = 0.285\ C\ d^{2.63}\ S^{0.54}$$

$$Q = (0.285)(140)(10)^{2.63}\ (0.0123)^{0.54} = 1589 \text{ gpm}$$

Storage Tank to Hospital

$$\text{Head Loss} = \frac{200 - 142.4}{4000} = 0.0144 \text{ ft/ft}$$

$$Q = (0.285)(140)(8)^{2.63}(0.0144)^{0.54} = 958 \text{ gpm}$$

Total flow rate available = 1589 gpm + 958 gpm = 2547 gpm

(c) $$\text{Hp} = \frac{QH}{3960 \times \text{Efficiency}}$$

$$\text{Hp} = \frac{1589 \text{ gpm} \times (173.25 \text{ ft} - 142.4 \text{ ft})}{0.72 \times 3960}$$

Hp = 17.2

Power = 17.2 Hp x 0.745 Kw/Hp = 12.8 Kw

PROBLEM 1.20

Reservoir A is at elevation 100 feet. From Reservoir A a 48 inch diameter pipe 5500 feet long discharging 50 cubic feet/second leads to a wye. From the wye it branches into two pipes: a 30 inch pipe 4500 feet long leading to Reservoir B at elevation 86 feet, and another pipe 4500 feet long leading to Reservoir C at elevation 80 feet. Calculate the diameter of the pipe leading to Reservoir C.

Solution

Solve using Manning equation. Assume n = 0.011.

From Manning equation, head loss for 48-inch pipe flowing full = 0.00085 ft/ft.

0.00085 ft/ft x 5500 ft = 4.7 ft head loss

Calculate head loss and flow for 30 inch pipe to Reservoir B.

Head Loss = (100 - 86) - 4.7 = 9.3 feet = 0.00207 ft/ft

Q = 21.5 cubic feet/second from Manning equation.

For pipe to Reservoir C,

$$Q = 50 - 21.5 = 28.5 \text{ cubic feet/second}$$

Head Loss = (100 - 80) - 4.7 = 15.3 ft = 0.0034 ft/ft

From Manning equation, diameter = 30 inches

PROBLEM 1.21

Estimate the cost of pumping one million gallons of water (68°) at a rate of 35 gallons per minute against a static head of 200 feet through new 6 inch cast iron pipe 1200 feet long. Pump efficiency is 80% and motor efficiency is 70%. Electric cost is 6 mills per kilowatt hour.

Solution

Calculate friction head loss from Hazen-Williams equation.

Assume C = 130.

$$Q = 0.285\ C\ d^{2.63}\ S^{0.54}$$

$$35 \text{ gpm} = (0.285)(130)(6)^{2.63}\ S^{0.54}$$

$$S = 0.000147 \text{ ft/ft}$$

Friction loss = 0.000147 ft/ft x 1200 ft = 0.177 ft

Total head loss = 200 + 0.177 ft = 200.177 ft

$$\text{Pump Hp} = \frac{H\ Q}{3960 \times \text{Efficiency}} = \frac{200.177 \text{ ft} \times 35 \text{ gpm}}{3960 \times 0.8 \times 0.7} = 3.16 \text{ hp}$$

$$\text{Pumping Cost} = 3.16 \text{ Hp} \times 0.746\ \frac{\text{kw}}{\text{Hp}} \times \frac{\$0.006}{\text{kw-hr}} \times \frac{10^6 \text{ gal}}{35 \text{ gpm} \times 60}$$

$$= \$6.74$$

PROBLEM 1.22

A sharp crested rectangular weir 5 feet high extends across a channel 10 feet wide. The head on the weir is 1.5 feet. Compute the discharge in cubic feet per second. The velocity of approach may be neglected.

Solution

Assume there are no end contractions.

Solve using Francis formula,

$$Q = 3.33\ L\ H^{3/2}$$

where Q = discharge in cfs

L = length of crest of weir, feet

H = head on weir, feet

$$Q = 3.33 \times 10 \text{ feet} \times (1.5)^{3/2} = 61.2 \text{ cfs}$$

PROBLEM 1.23

A storm sewer is to serve drainage basins I through VI as shown in Figure 1.5. Calculate size of each sewer based on the information in Table 1.6. Assume rainfall intensity = $\dfrac{60}{(t + 15)^{0.8}}$

Table 1.6

Data for Problem 1.23

Drainage Basin	Area Acres	Runoff Coeff	Inlet Time Minutes
I	3	0.4	10
II	3	0.4	10
III	4	0.4	12
IV	5	0.4	15
V	5	0.5	15
VI	6	0.5	20

Sewer	Length, feet	Slope, ft./ft.
AB	600	.008
BC	750	.0095
CD	800	.0095

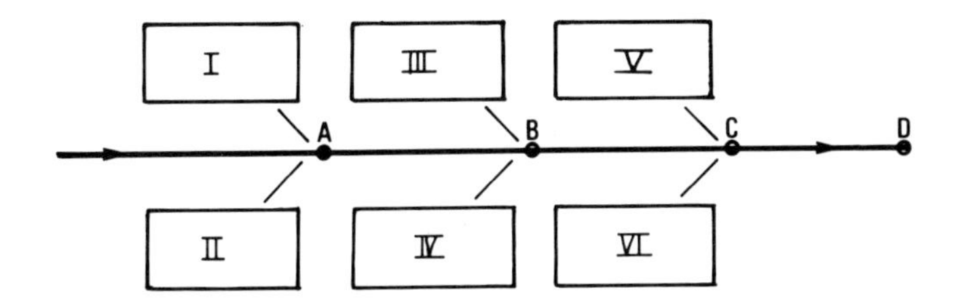

Figure 1.5. Drainage area for Problem 1.23.

Solution

The storm runoff can be estimated from the rational equation

$$Q = C\ I\ A$$

where Q = peak runoff rate, cfs

I = rainfall intensity, inches/hour

A = drainage area, acres

The required pipe diameter is calculated from the Manning equation with n = 0.013.

The inlet time for several drainage areas served by a sewer is taken as the maximum inlet time. Rainfall intensity is calculated from the equation

$$i = \frac{60}{(t + 15)^{0.8}}$$

with t = duration of rainfall in minutes
$t = t_c$, time of concentration of the area drained

Tabulated results are shown in Table 1.7.

The actual time of concentration is equal to the sum of the drainage inlet time and the flow time of the upstream sewer. Adjustment of computed values to account for this difference involves a trial-and-error solution.

Table 1.7

Results for Problem 1.23.

Sewer	Total Drainage Area	Inlet Time	t_c	i
AB	6 acres	10 min.	10 min.	4.59
BC	15 acres	15 min.	15 min.	3.95
CD	26 acres	20 min.	20 min.	3.49

Sewer	Discharge	Nominal Pipe Diameter
AB	11.0 cfs	21-inch
BC	21.5 cfs	24-inch
CD	40.1 cfs	30-inch

To determine whether this adjustment would be significant for this particular case, corrected values with consideration for upstream sewer time of flow are calculated for an initial velocity of 5.5 ft/sec in sewer AB and shown in Table 1.8. No change in sewer diameters previously calculated would appear to be necessary in this example.

Table 1.8

Corrected Results for Problem 1.23.

Sewer	Total Drainage Area	t_c	i
AB	6 acres	10 minutes	4.59
BC	15 acres	16.8 minutes	3.77
CD	26 acres	21.9 minutes	3.35

Sewer	Discharge	Velocity (flowing full)	Time of Flow	Nominal Pipe Diameter
AB	11.0 cfs	5.5 ft/sec	1.8 min.	21-inch
BC	22.6 cfs	7.2 ft/sec	1.9 min.	24-inch
CD	38.5 cfs	7.8 ft/sec	1.7 min.	30-inch

PROBLEM 1.24

A lake 50 feet deep is aerated with a two inch diameter diffuser pipe 450 feet long. Air flows at a rate of 200 standard cubic feet per minute (scfm). The pressure drop across the length of the diffuser is estimated as 3 psi. If an air compressor is located 300 feet from the lake, calculate the horsepower required for the compressor.

Solution

Calculate discharge pressure for diffuser.

Assume air transfer line = 300 ft + 50 ft = 350 feet

Head loss for air flow through 2-inch diameter steel pipe is approximately 1.9 psi/100 feet.

350 ft x 1.9 psi/100 ft = 6.65 psi

Add 10% for fittings = 0.67 psi

7.32 psi

Pressure at bottom of lake

50 ft x 14.7 psi/34 ft = 21.62 psi

Pressure drop through diffuser = 3.00 psi

Total discharge pressure required by compressor = 31.94 psi

Say 35 psi for design.

Assume adiabatic compression. Compressor horsepower and pressure, volume, and temperature relationships can be calculated from the following equations[10]

$$Hp = 0.00436\ Q\ p_1 \left[\frac{k}{k-1}\right] \left[\left(\frac{p_2}{p_1}\right)^{(k-1)/k} - 1\right]$$

where Hp = adiabatic horsepower

Q = air flow rate, cfm

k = constant = 1.395

p = pressure, psi

$$\frac{T_2}{T_1} = \left[\frac{V_1}{V_2}\right]^{k-1} \qquad \frac{T_2}{T_1} = \left[\frac{p_2}{p_1}\right]^{k-1/k} \qquad \frac{p_2}{p_1} = \left[\frac{V_1}{V_2}\right]^{k}$$

where T = temperature, degrees Rankine

V = volume, cubic feet

p = pressure, psi

$$\frac{T_2}{T_1} = \left[\frac{p_2}{p_1}\right]^{k-1/k}$$

Assume T_1 = temperature at compressor = 70° F = 530° R.

Calculate temperature T_2 at bottom of lake.

$$\frac{T_2}{T_1} = \left[\frac{p_2}{p_1}\right]^{k-1/k}$$

p_1 at compressor is assumed equal to 14.7 psi.

$$T_2 = (530^\circ\ R)\left[\frac{35\ psi}{14.7\ psi}\right]^{0.283} = 677.5^\circ\ R$$

Convert air flow rate of 200 scfm to air flow at 35 psi and 677.5° R. Standard temperature and pressure is taken as 32° F, 14.7 psi.

$$V_2 = \frac{14.7\ psi}{35.0\ psi} \times \frac{677.5^\circ\ R}{492^\circ\ R} \times 200\ cfm$$

$$V_2 = 116\ cfm$$

$$\left[\frac{V_1}{116}\right]^{1.395} = \frac{35}{14.7}$$

$$V_1 = 216\ cfm$$

Calculate compressor horsepower.

$$Hp = 0.00436 \times 216 \times 14.7 \times \left[\frac{1.395}{1.395 - 1}\right]\left[\frac{35}{14.7}^{0.283} - 1\right]$$

$$Hp = 13.6$$

PROBLEM 1.25

Sulfuric acid (specific gravity = 1.83) is to be pumped through a 6-inch line at the rate of 650 gallons per minute by a 20 horsepower pump. The pipe discharges into a closed tank. The acid level in the tank is 50 feet above the intake pipe which may be assumed to be at atmospheric pressure. Head loss in the suction line is 2 feet and in the discharge line 6 feet.

(a) What is the pressure on the surface of the acid in the closed tank?

(b) What pressure would be indicated by a gauge on the discharge side of the pump?

Solution

(a) $Q = 650 \text{ gpm} = 1.448 \text{ ft}^3/\text{sec}$

$$\text{Pump energy, } W_p = \frac{20 \text{ hp} \times 550 \text{ ft-lb/sec}}{62.4 \text{ lb/ft}^3 \times 1.83 \times 1.448 \text{ ft}^3/\text{sec}}$$

$$W_p = 66.53 \text{ feet}$$

Identify point 1 as suction line, point 2 as discharge line and point 3 as tank surface.

Write Bernoulli equation between points 1 and 3.

$$p_1 + \frac{V_1^2}{2g} + Z_1 + W_p = p_3 + \frac{V_3^2}{2g} + Z_3 + h_f$$

$$p_1 = \frac{14.7 \text{ lb/in}^2 \times 144 \text{ in}^2/\text{ft}^2}{62.4 \text{ lb/ft}^3 \times 1.83} = 18.54 \text{ feet}$$

$$18.54 + \frac{(7.39)^2}{64.4} + 0 + 66.53 = \frac{p_3 \times 144}{62.4 \times 1.83} + 0 + 50 + 8$$

$$p_3 = 22.14 \text{ psi} = 7.44 \text{ psi gauge}$$

(b) Write Bernoulli equation between points 1 and 2.

$$18.54 + \frac{(7.39)^2}{64.4} + 0 + 66.53 = p_2 + \frac{(7.39)^2}{64.4} + 0 + 2$$

p_2 = 65.87 psi = 51.17 psi gauge

PROBLEM 1.26

An existing single-pipe inverted siphon is to be replaced due to operational problems caused by sedimentation. The required length is 170 feet, available fall (invert to invert) is 2.25 feet, and the maximum depression of the siphon is 9.0 feet. Determine pipe size required for an average daily flow of 1200 gallons per minute and a peak flow of 4 times average daily flow. Assume inlet loss equals 0.4 feet and pipe velocity is 3 feet/second.

Solution

A two-barrel siphon is recommended to ensure adequate scour velocity is maintained under varying flow conditions. Assume that cast iron pipe is used.

1200 gpm = 1.73 mgd

4800 gpm = 6.92 mgd

$$\text{Slope} = \frac{2.25 - 0.40}{170} = 10.88 \text{ ft/1000 ft.}$$

Two pipes are required. Determine pipe size using Hazen-Williams equation with C = 100.

Required capacity of first pipe is 1.73 mgd. A 10-inch pipe will carry only 1.5 mgd. Choose a 12-inch pipe to carry 2.42 mgd at a velocity of 4.8 ft/sec when flowing full.

Second pipe must have a capacity of 6.92 - 2.42 = 4.50 mgd

Choose a 15-inch pipe to carry 4.5 mgd with velocity of 5.5 ft/sec.

Two pipes are required. A 12-inch and 15-inch.

References

1. U.S. Department of Commerce, Bureau of Public Roads, Design Charts for Open-Channel Flow, page 100, August, 1961.

2. Ingersoll-Rand Company, Woodcliff Lake, N.J., Condensed Hydraulic Data, page 6, 1973.

3. Moody, L.F., "Friction Factors for Pipe Flow," Transactions ASME, 66:671, 1944.

4. Crane Company, "Flow of Fluids," Technical Paper No. 410, page 3-2, 1969.

5. Gallant, Robert W., "Sizing Pipe for Liquids and Vapors," Chemical Engineering, February 24, 1969, p. 104.

6. Stindt, William H., "Pump Selection," Chemical Engineering, October 11, 1971, p. 44.

7. Op. Cit., Ingersoll-Rand Company, Condensed Hydraulic Data.

8. Epsey, W.H. and D.E. Winslow, The Effects of Urbanization on Unit Hydrographs for Small Watersheds, 1968 in Water Quality Management Planning for Urban Runoff, U.S. Environmental Protection Agency Report No. EPA - 440/9-75-004, December, 1974.

9. Fair, Gordon Maskew, J.C. Geyer, and D.A. Okun, Water and Wastewater Engineering, John Wiley & Sons, Inc., New York, 1966, p. 14-13.

10. Perry, Robert H., Cecil H. Chilton, Sidney D. Kirkpatrick, Editors, Chemical Engineers' Handbook, Fourth Edition, McGraw-Hill Book Company, New York, 1963, pp. 6-15, 6-16.

Additional References

1. Water Pollution Control Federation Manual of Practice No. 8, _Sewage Treatment Plant Design_, 1959.

2. Symons, George E., _Pumps and Pumping_, Water and Wastes Engineering/Manual of Practice Number One, Dun-Donnelley Publishing Corporation, New York, 1966.

3. King, Horace W., Chester O. Wisler, James G. Woodburn, _Hydraulics_, John Wiley and Sons, Inc., New York, 1958.

4. Water Pollution Control Federation Manual of Practice No. 9, _Design and Construction of Sanitary and Storm Sewers_, 1970.

Chapter 2

WATER SUPPLY AND TREATMENT

PROBLEM 2.1

A water treatment plant is to treat 25 million gallons per day of water from a large river. Treatment consists of rapid mixing, flocculation, sedimentation, standard rate sand filtration, and chlorination. Alum is added at the rate of 1.5 grains per gallon. Peak daily demand is 220% of average daily flow. Calculate the following:

(a) Design capacity and size of each treatment unit.

(b) The amount of alum and chlorine required on a daily basis.

Solution

(a) Flash Mixing

Design for minimum detention period of 30 seconds.

$$\text{Tank Capacity} = \frac{25 \times 10^6 \text{ gal}}{24 \text{ hrs} \times 60} \times 0.5 \text{ min} \times \frac{\text{ft}^3}{7.48 \text{ gal}}$$

$$= 1160 \text{ ft}^3$$

For square tank 10 feet deep, Area = 116 ft^2

Tank Dimensions: 10.5 ft x 10.5 ft x 10 ft deep

Flocculation

Use horizontal paddle mixer in flocculation basins. Design for 30 minutes detention time (minimum required).

$$\text{Tank Capacity} = \frac{25 \times 10^6 \text{ gal}}{24 \text{ hrs}} \times 0.5 \text{ hr} \times \frac{\text{ft}^3}{7.48 \text{ gal}}$$

For 15 foot deep basin, Area = 4642 sq. ft.

Choose four basins each 1161 sq. ft.

Assume length: width ratio of 2:1 for rectangular tank.

Tank Dimensions: 15 ft deep x 24 ft x 48 ft

Sedimentation

Design for four hour detention time, loading rate of 360 - 540 gpd/sq. ft. surface area.

$$\text{Tank Capacity} = \frac{25 \times 10^6 \text{ gal}}{24 \text{ hrs}} \times 4 \text{ hrs} \times \frac{\text{ft}^3}{7.48 \text{ gal}}$$

$$= 557{,}000 \text{ ft}^3$$

For 12 foot tank depth,

Total Area = 46,420 ft^2

Surface loading = 538 gpd/ft^2

Use eight basins, each 5803 ft^2.

Assume length: width ratio of 4:1 for rectangular basin.

Tank Dimensions: 12 ft deep x 38 ft x 152 ft

Filtration

Design for filtration rate of 2 gpm/ft^2.

Assuming each filter unit will treat 2.5 mgd, 10 beds will be required.

Assume 5 filters are backwashed for 20 minutes each day at a rate of 18 gpm/sq. ft. Also provide for surface wash. Total backwash time is 100 minutes.

Calculate required filter surface area.

$$\frac{25 \times 10^6 \text{ gal/day}}{2 \text{ gpm/ft}^2 \times 60 \text{ min/hr} \times (24 \text{ hr} - 100/60 \text{ hr})} = 9330 \text{ ft}^2$$

$$= 933 \text{ ft}^2\text{/filter}$$

Assume length: width ratio of 2:1.

Filter Dimension: 21.5 ft x 43 ft

(b) Calculate alum required at dosage rate of 0.5 grains/gallon.

1 grain/gallon = 142.5 lbs/million gallons

1.5 x 142.5 x 25 = 5344 lbs/day alum

Calculate chlorine required. Assume maximum dosage of 10 mg/l.

10 mg/l x 8.34 x 25 mgd = 2085 lbs/day

Design for chlorination capacity equal to 150 % of maximum demand or 3128 lbs/day.

PROBLEM 2.2

Using the data provided in Problem 2.1, calculate the size of a distribution reservoir required for fire protection.

Solution

Assume water consumption of 125 gallons/capita/day.

Using equation from the National Board of Fire Underwriters (American Insurance Association) fire flow demand can be estimated as,

$$Q = 1020 (P)^{\frac{1}{2}} [1 - 0.01 (P)^{\frac{1}{2}}]$$

where Q = required flow, gpm

P = Population in thousands

$$\text{Population} = \frac{25 \times 10^6 \text{ gal/day}}{125 \text{ gal/cap/day}} = 200,000$$

$$Q = 1020\ (200)^{\frac{1}{2}}\ [1 - 0.01\ (200)^{\frac{1}{2}}] = 12,390 \text{ gpm}$$

For a required fire flow of 10,000 gpm or greater, a 10 hour storage capacity is recommended.

12,390 gpm x 60 x 10 hr = 7,434,000 gallon storage for fire protection.

Provide storage for domestic use equivalent to 50% of peak daily demand. Peak demand is 220% of average daily flow.

0.5 (2.2 x 25 mg) = 27.5 million gallons

Total storage (domestic and fire) = 35 million gallons.

Alternately, calculate storage required for peak domestic use and fire use minus plant capacity during fire period.

(2.2 x 25 mg) + 7.434 mg - 10/24 (25mg) = 52 million gallons storage capacity.

PROBLEM 2.3

A water filtration plant has a design capacity of 5 mgd and has three multimedia filters. Previous data indicates that a filtration rate of 4 gpm/square foot can be used. Develop the preliminary design for a filter underdrain system if the rate of backwash will be 18 gpm/square foot.

Solution

(a) Determine filter dimensions.

Required surface area = $(5\text{mgd}/1440)/(4 \text{ gpm/ft}^2)$
$= 867 \text{ ft}^2 = 289 \text{ ft}^2\text{/filter}$

Assuming filter length is twice width,

$12 \text{ ft} \times 24 \text{ ft} = 288 \text{ ft}^2$

Use filter slightly oversized or 13 ft x 24 ft to account for out of service time for backwashing. Filter area equals 312 square feet.

(b) Determine size and spacing for filter laterals, orifices, and center manifold based on backwash water requirements.

Assume total orifice area = 0.2% - 0.5% bed area

$= 0.62 - 1.56\ ft^2$ per filter

Using ½ inch orifice opening, area = $0.00136\ ft^2$

Calculate minimum and maximum number of orifices required per filter.

Minimum = $0.62\ ft^2/0.00136\ ft^2 = 458$

Maximum = $1.56\ ft^2/0.00136\ ft^2 = 1147$

Use laterals on 9-inch centers extending from center manifold and orifices on 6-inch centers on laterals. Calculate approximate number of orifices as,

64 laterals/filter x 13 orifices/lateral = 832

Ratio of total lateral area/total orifice area will be in the range of 2:1 to 4:1, equivalent to a range of 4.7 sq. in. to 9.4 sq. in. Lateral diameter must therefore be 2.4 inches to 3.5 inches. Select 3-inch lateral.

Check velocity in lateral.

Backwash rate = $312\ ft^2 \times 18\ gpm/ft^2$

= 5616 gpm = 12.5 cfs

12.5 cfs/64 = 0.195 cfs/lateral = 4 ft/sec

Velocity of backwash water through manifold should not exceed approximately 6 - 10 ft/sec.

For 16-inch diameter pipe,

Velocity = $12.5\ cfs/1.39\ ft^2 = 9.0$ ft/sec

PROBLEM 2.4

A variable rate filter is designed for an average filter rate of 6 gpm/ft^2. The filter media consists of 30 inches of sand with an effective diameter of 0.55 mm. The underdrain pipe losses are estimated to be 3 feet at the average filter rate.

(a) Discuss the principle of operation of a variable rate as opposed to a constant rate filter.

(b) Calculate the head loss through the clean media.

(c) Determine the depth of water over the sand media at the beginning of the filter run.

Solution

(a) Variable rate filtration is also called declining rate filtration. With variable rate filtration a constant head loss is maintained across the filter and clearwell to control the flow rate. The rate of filtration is at a maximum at the beginning of the filter run but decreases as the filter becomes plugged and head loss increases.

In a constant rate filter the flow rate is held constant through the filter run by a rate control valve. The rate controller opens more as the head loss through the filter increases during the filter run to maintain a constant rate of flow.

(b) Calculate head loss through clean sand using equation of Fair and Hatch[1] as given by Camp,[2]

$$S = \frac{h}{\ell} = \frac{\beta \nu}{g} \frac{(1 - p)^2}{p^3} \frac{s^2}{d^2} V$$

where S = head loss (ft) through sand bed thickness, ℓ (feet)

β = a constant

ν = Kinematic viscosity, ft^2/sec

g = acceleration of gravity, ft/sec^2

p = porosity ratio

s = shape factor

d = effective diameter of sand, feet

V = average velocity, gpm/ft^2

For spherical sand particles use,

$s = 6.0$

$\beta = 5.0$

$p = 0.4$

$\nu = 1.0105$ centistokes x 1.075×10^{-5} =
1.086×10^{-5} ft^2/sec

$d = 0.55$ mm = 0.0018 feet

$V = 6$ gpm/ft^2 = 0.0134 ft/sec

$$S = \frac{5 \times 1.086 \times 10^{-5}}{32.2} \times \frac{(1 - 0.4)^2}{(0.4)^3} \times \frac{(6.0)^2}{(0.0018)^2} \times 0.0134$$

$S = 1.4$ ft/ft

$S = 3.5$ ft water for 30 inch bed depth

(c) Pipe losses at 6 gpm = 3 feet

Assume maximum filter rate = 6 gpm/ft^2 x 1.33 = 8 gpm/ft^2

Calculate pipe loss at 8 gpm/sq. ft. Assume pipe loss is proportional to square of filter rate.

$(8)^2/(6)^2 \times 3$ ft = 5.3 feet

Total head loss at 8 gpm = head loss (piping)
+ head loss (sand)
= 8.8 feet

Design for 9 foot water depth, depending on operating conditions.

PROBLEM 2.5

An ion exchange process is to be used to soften water with an analysis as shown in Table 1. A synthetic zeolite resin is to be used with an exchange capacity of 20 Kilograins of $CaCO_3$ per cubic foot when regenerated at the rate of 15 pounds of salt per cubic foot. Flow rate is 300 gpm. Assume a final hardness level of 2 ppm $CaCO_3$ is required.

(a) List several of the advantages and disadvantages of the ion exchange process for water softening.

(b) Write the general chemical reactions that occur in the ion exchange process.

(c) Calculate the bed depth and shell diameter for the ion exchange equipment.

(d) Calculate the pounds of salt required per month for zeolite regeneration.

Table 2.1

WATER ANALYSIS FOR PROBLEM 2.5.

Component	mg/l as $CaCO_3$
Calcium	85
Magnesium	26
Sodium	37
Chloride	10
Sulfate	60
Bicarbonate	78

Solution

(a) Several advantages and disadvantages of an ion exchange system are as follows:

Advantages	Disadvantages
Properly designed systems are relatively easy to operate.	Reduction in alkalinity or total solids can not be achieved.
Near zero hardness levels can be achieved in most cases.	Waste regenerant brine may cause disposal problems.
Regeneration of ion resin with common salt can be readily accomplished.	Relatively high sodium level in treated water may cause health problem.

(b) Zeolites are used in the ion exchange process for water softening to absorb calcium and magnesium (the principle elements that cause water hardness). The calcium and magnesium ions are exchanged for sodium ions. Using Z for the zeolite compound the basic reactions are written in the ionic form,

$$Na^{+2}Z + Ca^{+2} \longrightarrow Ca^{+2}Z + 2Na^{+}$$

$$Na^{+2}Z + Mg^{+2} \longrightarrow Mg^{+2}Z + 2Na^{+}$$

Other ions may be substituted for calcium and magnesium in the reaction.

If calcium or magnesium is in the form of a bicarbonate or sulfate, the equation may be written in molecular form as,

$$Ca(HCO_3)_2 + Na_2Z \longrightarrow CaZ + 2Na(HCO_3)$$

$$CaSO_4 + Na_2Z \longrightarrow CaZ + Na_2SO_4$$

The same reactions occur with magnesium.

After the zeolite bed is exhausted it may be regenerated with sodium chloride according to the equation,

$$2NaCl + CaZ \longrightarrow Na_2Z + CaCl_2$$

$$2NaCl + MgZ \longrightarrow Na_2Z + MgCl_2$$

(c) Use two units and assume a 24 hour regeneration cycle for each softener unit. Actual frequency of regeneration would depend upon operating conditions.

For water analysis, hardness = calcium + magnesium ions = 111 mg/l

$$\text{Hardness to be removed} = \frac{(111 - 2)\ \text{mg/l as } CaCO_3}{17.1\ \text{mg/l per grain/gallon}}$$

$$= 6.37\ \text{grains/gallon}$$

Calculate hardness removed in 24 hours.

$$300\ \frac{\text{gal}}{\text{min}} \times 6.37\ \frac{\text{grains}}{\text{gal}} \times \frac{1440\ \text{min}}{\text{day}} = 2752\ \text{Kilograins/day}$$

$$\text{Resin required} = \frac{2752 \text{ Kgr}}{20 \text{ Kgr/ft}^3} = 138 \text{ ft}^3$$

Design for a maximum flow rate of 5 gpm/cu.ft. of resin.

Bed depth should not be less than approximately 30 inches or greater than 72 inches.

Using two units requires 69 ft^3 resin/unit.

Maximum flow rate through one unit is,

$$\frac{300 \text{ gpm}}{69 \text{ ft}^3} = 4.34 \text{ gpm/ft}^3$$

Using a 60 inch diameter shell provides an area of 19.6 square feet.

$$\text{Bed depth} = \frac{69 \text{ ft}^3}{19.6 \text{ ft}^2} = 3.52 \text{ feet} = 42 \text{ inches}$$

Calculate flow rate through each unit at nominal flow.

$$\frac{150 \text{ gpm}}{69 \text{ ft}^3} = 2.2 \text{ gpm/ft}^3$$

(d) For zeolite regeneration,

$$\frac{15 \text{ lbs salt}}{\text{ft}^3 \text{ resin}} \times 138 \text{ ft}^3 = 2070 \text{ lbs salt/day}$$

$$= 62{,}100 \text{ lbs/month}$$

PROBLEM 2.6

One million gallons of raw water containing 120 mg/l suspended solids is clarified by sedimentation. The clarifier effluent has a suspended solids concentration of 5.0 mg/l and the clarifier sludge a 3.5% solids concentration by weight.

(a) How many pounds of dry solids are removed per million gallons of raw water.

(b) Calculate the flow rate of the sludge in gallons and pounds per day.

(c) If the sludge stream is filtered to produce a 0.4% filtrate and a wet cake containing 20% solids, calculate the volume of the filtrate and pounds of wet cake produced.

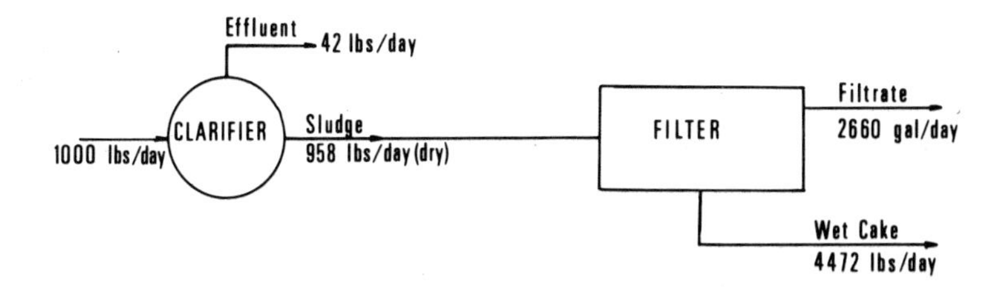

Figure 2.1. Flow diagram for Problem 2.6.

Solution

(a) Assume density of water = 8.34 lbs/gallon.

Write a mass balance around the clarifier:
Dry solids in = Dry solids out.

Dry solids in = 1 mgd x 8.34 lb/gal x 120 mg/l
= 1000 lbs/day

Dry solids in effluent = 1 mgd x 8.34 x 5 mg/l
= 42 lbs/day

Dry solids removed in sludge = 1000 - 42
= 958 lbs/million gallons treated.

(b) Assume density of sludge = 8.6 lbs/gallon

Sludge is 3.5% by weight dry solids.

958 lbs/0.035 = 27,400 lbs/day wet cake

$$\frac{27{,}400 \text{ lbs}}{8.6 \text{ lbs/gal}} = 3180 \text{ gallons}$$

(c) Write a mass balance around filter. Let X = flow rate of filtrate in gallons.

958 lbs = 8.34 lbs/gal x (X) gal x 0.004 + (3180 - X) x 8.6 lbs/gal x 0.2

X = 2660 gallons filtrate

At a density of 8.6 lbs/gal,

(3180 - 2660) x 8.6 lbs/gal = 4472 lbs wet cake

PROBLEM 2.7

Calculate the number of quarts of sodium hypochlorite solution required to disinfect a well casing 6 inches in diameter and 150 feet deep. The sodium hypochlorite solution is available as 5.25% chlorine. Considering that disinfection is the final step in completion of a well, describe briefly the disinfection procedure using sodium hypochlorite solution.

Solution

A minimum chlorine concentration of 100 mg/l is recommended. Calculate the volume of water to be treated.

Area of well casing $= \frac{\pi \, (\frac{1}{2})^2}{4} = 0.196 \text{ ft}^2$

Volume water $= 0.196 \text{ ft}^2 \times 150 \text{ ft} \times 7.48 \text{ gal/ft}^3 = 220$ gallons

Weight chlorine per gallon $= 8.34 \times 0.0525 = 0.42$ lbs NaOCl solution

Calculate pounds chlorine required at 100 mg/l concentration.

$100 \text{ mg/l} \times 220 \text{ gal} \times 8.34 \text{ lb/gal} = 0.1835$ lbs

Gallons NaOCl solution $= \frac{0.1835 \text{ lb}}{0.42 \text{ lb/gal}} = 0.44$ gallons or 2 quarts

The completed well should initially be cleaned as completely as possible to remove oil, grease, and dirt from the well prior to disinfection. The required amount of disinfectant should be added, mixed well, and left in the well for at least 12 hours. The well is then pumped long enough to clear it of the chlorine.

PROBLEM 2.8

Analysis of a raw water in a treatment process indicates a pH of 7.4 and alkalinity (as $CaCO_3$) of 120 mg/l. It is desired to lower the pH to 6.7 for coagulation by adding 1.9 grains of alum per gallon water at a cost of $5.00 per 100 pounds, or by adding only 1.0 grains per gallon alum and an equivalent amount of 66° Be′ sulfuric acid that can be purchased for 3¢ per pound. For each million gallons of water treated, calculate the following:

(a) How much natural alkalinity will remain after treatment?

(b) Determine the theoretical amount of sludge produced by adding 1.9 grains of alum.

(c) Which is the cheaper method of treatment?

Solution

120 mg/l alkalinity (as $CaCO_3$) = 2.4 meq/l

Theoretical reaction can be written as,

$$Al_2(SO_4)_3 \cdot 14.3\ H_2O + 3\ Ca(HCO_3)_2 = 3\ CaSO_4 + 2\ Al(OH)_3 + 6\ CO_2 + 14.3\ H_2O$$

Filter Alum $[Al_2(SO_4)_3 \cdot 14.3H_2O]$ — Molecular Weight = 600; Equivalent Weight = 100

$Ca(HCO_3)_2$ — Molecular Weight = 162; Equivalent Weight = 81

$Al(OH)_3$ — Molecular Weight = 78; Equivalent Weight = 26

1.0 grain/gallon = 142.5 lbs/million gallons
= 17.1 mg/l

Alum added = 142.5 x 1.9 = 271 pounds
= 17.1 x 1.9 = 32.5 mg/l

271 lbs/(600lbs/lb-mole) = 0.45 lb-mole alum

(a) 1.0 mg/l alum reacts with 0.5 mg/l natural alkalinity.

32.5 mg/l alum x 0.5 = 16.25 mg/l natural alkalinity destroyed

Alkalinity remaining = 120 - 16.25 = 103.75 mg/l

(b) Based on balanced equation above, 0.45 lb-moles alum will theoretically produce 0.90 lb-moles of aluminum hydroxide [$Al(OH)_3$] floc or sludge.

0.9 lb-moles x 78 lb/lb-mole = 70 lbs sludge

(c) Calculate comparative cost of each treatment scheme for one million gallons of water treated.

Alum addition only.

Cost = 271 lbs x $5.00/100 lbs = $13.55

Alum and acid addition.

Alum added = 142.5 lbs = 17.1 mg/l

Alkalinity destroyed = 17.1 x 0.5 = 8.55 mg/l

Alkalinity remaining to be destroyed by acid. = 16.25 - 8.55 = 7.7 mg/l

105 lbs 66° H_2SO_4 will neutralize 100 mg/l alkalinity.

105 lbs/100 mg/l x 7.7 mg/l = 8.09 pounds acid

Alum cost = 142 lbs x $5.00/100 lbs	= $7.10
Acid cost = 8.09 lbs x $0.03/lb	= 0.24
Total	$7.34

Alum and acid solution will therefore result in a savings of $6.21 for each million gallons of water treated.

PROBLEM 2.9

A new residential subdivision is planned as shown in Figure 2.2. Fire hydrants with two 2½ inch outlets are to be installed. On the basis of zoning and comparison with similar areas in the city, fire flow demand is 2000 gallons per minute. Show the typical location of hydrants for the area.

In testing the system a residual pressure of 57 psig is measured. During the test a hydrant with one nozzle open and discharging the same amount of flow measured a pitot pressure drop of 13 psig. Determine the rate of flow available at a residual pressure of 20 psi.

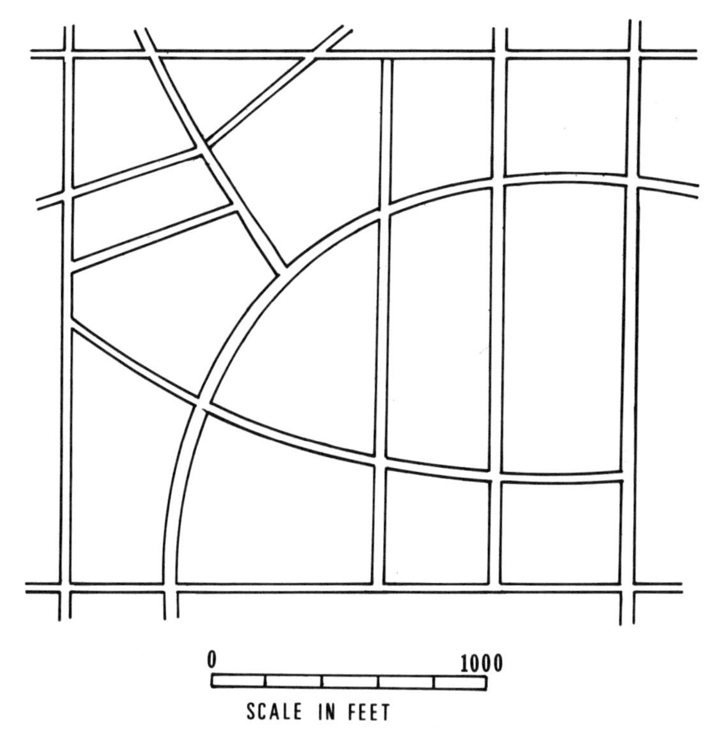

Figure 2.2. Street plan for Problem 2.9.

Solution

In residential areas hydrants are usually located so that the distance between hydrants does not exceed 500 feet. Figure 2.3 shows a possible hydrant distribution for the area. Thirty-two hydrants would be installed. The service area for this problem is approximately 3,600,000 square feet. The area served per hydrant is 112,500 square feet which should be adequate under most circumstances for fire protection with a fire flow demand of 2000 gpm.

The discharge rate for a hydrant is calculated from the equation,

$$Q = 29.82\ Cd^2\ p^{\frac{1}{2}}$$

where Q = Flow, gallons/minute

C = coefficient of discharge

d = hydrant nozzle diameter, inches

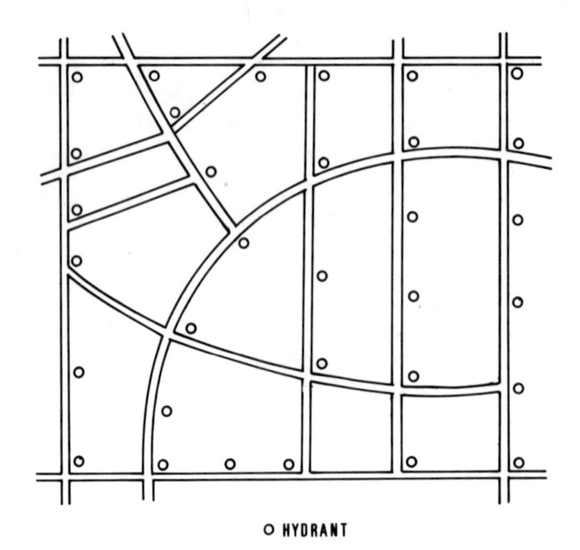

Figure 2.3. Fire hydrant location for Problem 2.9.

p = pitot reading, psi

For C = 0.9 the equation reduces to

$$Q = 26.8\ d^2\ p^{\frac{1}{2}}$$

The actual discharge during the test may be calculated as,

$$Q = 26.8\ (2.5)^2\ (13)^{\frac{1}{2}} = 604\ \text{gpm}$$

The rate of flow at a residual pressure of 20 psi can be calculated using the equation,

$$Q_R = Q_F\ \frac{H_R^{0.54}}{H_F^{0.54}}$$

where Q_R = discharge at specified residual pressure, gpm

Q_F = actual discharge during flow test, gpm

H_R = pressure drop from the original pressure to the specified residual pressure, psi

H_F = pressure drop during the flow test, psi

The residual pressure is the pressure in the water line when water is flowing through the hydrant nozzle. Static pressure is theoretically the pressure when no water is flowing.

Static pressure may be assumed equal to,

$$13 \text{ psi} + 57 \text{ psi} = 70 \text{ psi}$$

$$Q_R = 604 \text{ gpm} \frac{(70 \text{ psi} - 20 \text{ psi})^{0.54}}{(13 \text{ psi})^{0.54}}$$

$$Q_R = 1250 \text{ gpm}$$

PROBLEM 2.10

A surface water has the following analysis: Calcium 72.0 mg/l, Magnesium 48.8 mg/l, Sodium 9.2 mg/l, Bicarbonate 305 mg/l, Sulfate 134.4 mg/l, Chloride 7.1 mg/l.

(a) Calculate the number of milliequivalents per liter (meq/l) for each substance.

(b) Calculate the total hardness, carbonate and non-carbonate hardness, and alkalinity expressed as mg/l Ca CO_3.

(c) Draw a bar diagram of the water.

Solution

(a) Concentration in meq/l can be calculated by the equation,

$$\text{meq/l} = \frac{\text{mg/l}}{\text{equivalent weight}}$$

Concentration expressed as mg/l $CaCO_3$ can be calculated from the equation,

$$\text{mg/l } CaCO_3 = \text{mg/l} \times \frac{50}{\text{equivalent weight}}$$

Equivalent weights for a number of elements and compounds and compounds are listed in the Appendix. The data for this problem is summarized in Table 2.2.

Table 2.2

WATER ANALYSIS FOR PROBLEM 2.10.

Component	mg/l	Equivalent Weight	meq/l	mg/l as $CaCO_3$
Calcium Ca^{++}	72.0	20.0	3.6	180
Magnesium Mg^{++}	48.8	12.2	4.0	200
Sodium Na^{+}	9.2	23.0	0.4	20
			8.0	400
Bicarbonate HCO_3^-	305	61.0	5.0	250
Sulfate $SO_4^{=}$	134.4	48.0	2.8	140
Chloride Cl^-	7.1	35.5	0.2	10
			8.0	400

(b) Alkalinity represents the contents of carbonates, bicarbonates, hydroxides, and borates, silicates, and phosphates if present. Hardness represents the sum of multivalent metallic cations, in this case magnesium and calcium. When the total hardness exceeds the carbonate and bicarbonate alkalinity, the hardness equivalent to the alkalinity is carbonate hardness and the amount in excess of carbonate hardness is non-carbonate hardness. When the total hardness is equal to or less than the carbonate and bicarbonate alkalinity, then the total hardness is simply equivalent to the carbonate hardness. Non-carbonate hardness is zero.

For this analysis,

Alkalinity = Bicarbonate Alkalinity = 250 mg/l as $CaCO_3$

Total Hardness = 180 + 200 = 380 mg/l as $CaCO_3$

Carbonate Hardness = Alkalinity = 250 mg/l as $CaCO_3$

Non-carbonate Hardness = 380 - 250 = 130 mg/l as $CaCO_3$

(c) A bar diagram of the raw water is shown in Figure 2.4.

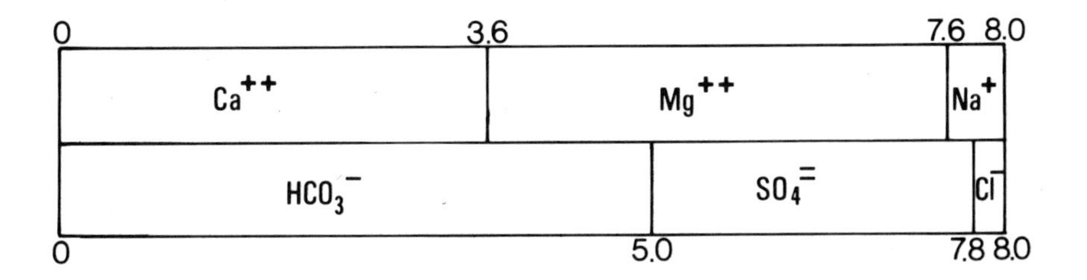

Figure 2.4. Bar diagram of water analysis for Problem 2.10.

PROBLEM 2.11

A raw water with an analysis as shown in Table 2.3 is to be treated using a lime soda-ash softening process. Excess lime and soda ash is added to achieve a residual hardness to the practical limit of 30 mg/l as $CaCO_3$ and 10 mg/l $Mg(OH)_2$ as $CaCO_3$.

(a) Discuss the chemical reactions that occur in the lime soda-ash softening process.

(b) List several advantages and disadvantages of the process.

(c) Calculate the quantity of chemicals required for softening and recarbonation.

Table 2.3

RAW WATER ANALYSIS FOR PROBLEM 2.11.

	mg/l	Equivalent Weight	meq/l	mg/l as $CaCO_3$
Calcium	80	20.0	4.0	200
Magnesium	30	12.2	2.5	125
Sodium	19	23.0	0.8	30
Chloride	18	35.5	0.5	25
Sulfate	64	48.0	1.3	67
Bicarbonate	336	61.0	5.5	275
Carbon dioxide (free)	15	22.0	0.7	35

Solution

(a) Basic reactions for the lime-soda ash water softening process using hydrated lime, $Ca(OH)_2$ are written as:

Lime

$$CO_2 + Ca(OH)_2 = CaCO_3 + H_2O$$

$$Ca(HCO_3)_2 + Ca(OH)_2 = 2CaCO_3 + 2H_2O$$

$$Mg(HCO_3)_2 + 2Ca(OH)_2 = 2CaCO_3 + Mg(OH)_2 + 2H_2O$$

In these reactions carbonate hardness in the form of calcium and magnesium bicarbonate is removed by reaction with lime to form insoluble $CaCO_3$ and $Mg(OH)_2$. Lime also reacts with free CO_2 in the water to form a $CaCO_3$ precipitate.

Soda Ash - Na_2CO_3

$$CaSO_4 + Na_2CO_3 = CaCO_3 + Na_2SO_4$$

$$MgSO_4 + Na_2CO_3 + Ca(OH)_2 = Mg(OH)_2 + CaCO_3 + Na_2SO_4$$

Noncarbonate hardness in the form of calcium and magnesium sulfates (or chlorides) is precipitated as $Mg(OH)_2$ and $CaCO_3$.

(b)

Advantages	Disadvantages
Large quantities of water can be softened economically.	Large quantities of chemical sludge are produced creating disposal problems.
Alkalinity and total solids are reduced. Water is low in color turbidity.	Complete softening to low hardness levels can not be achieved.
Disinfection is achieved by the effect of lime and high pH.	Chemical feed rates must be controlled closely.
Reduction in silica can achieved.	Operating problems may be experienced in filtration of softened water.

(c) A bar graph of the raw water is shown in Figure 2.5. The required amount of lime is calculated as,

$$\text{Lime} = \text{meq/l}\ [CO_2 + Ca(HCO_3)_2 + Mg(HCO_3)_2 + MgSO_4]$$

$$= 0.7 + 4.0 + 1.5 + 1.0$$

$$= 7.2\ \text{meq/l}$$

Equivalent weight of lime (CaO) = 28.0
Use 35 mg/l excess lime.
Lime dosage = (7.2 x 28.0) + 35 = 237 mg/l
= 1976 lbs/million gallons.

The required amount of soda ash is calculated as,

$$\text{meq/l}\ Na_2CO_3 = \text{meq/l}\ MgSO_4 = 1.0$$

Equivalent weight soda ash = 53.
Soda ash dosage = 1.0 x 53 = 53 mg/l
= 442 lbs/million gallons.

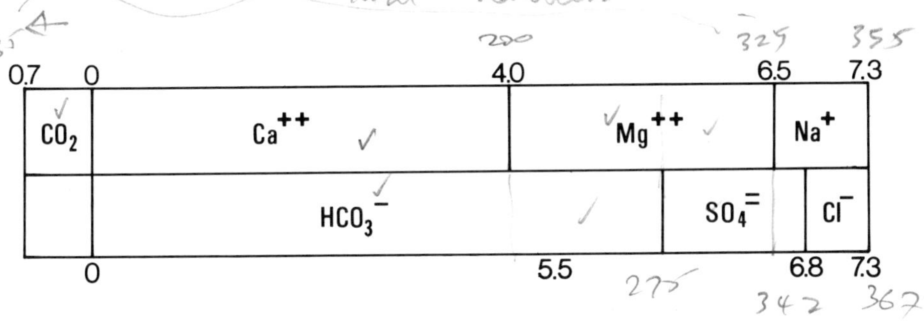

Figure 2.5. Raw water bar diagram. Problem 2.11.

Calculate carbon dioxide required for recarbonation. Carbon dioxide neutralizes excess lime and OH^- according to the equation,

$$Ca(OH)_2 + CO_2 = CaCO_3 + H_2O$$

Equivalent weight CO_2 = 22.0

Excess lime = 1.25 meq/l

OH^- = 0.20 meq/l

CO_2 = (1.25 + 0.20) x 22 = 31.9 mg/l

Additional carbon dioxide converts remaining alkalinity as carbonate ion to bicarbonate ion according to the equation,

$$CaCO_3 + CO_2 + H_2O = Ca(HCO_3)_2$$

Assuming all of remaining alkalinity is converted the bicarbonate form,

$$CO_2 = 0.6 \text{ meq/l} \times 22 = 13.2 \text{ mg/l}$$

$$\text{Total } CO_2 = 45.1 \text{ mg/l}$$

Bar diagrams of the softened water before recarbonation and the finished water are shown in Figure 2.6.

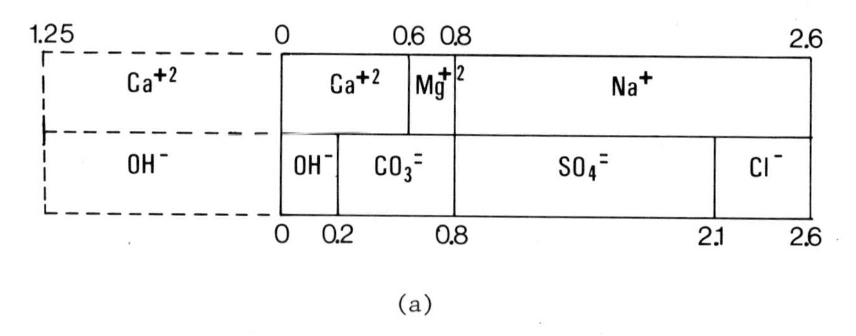

(a)

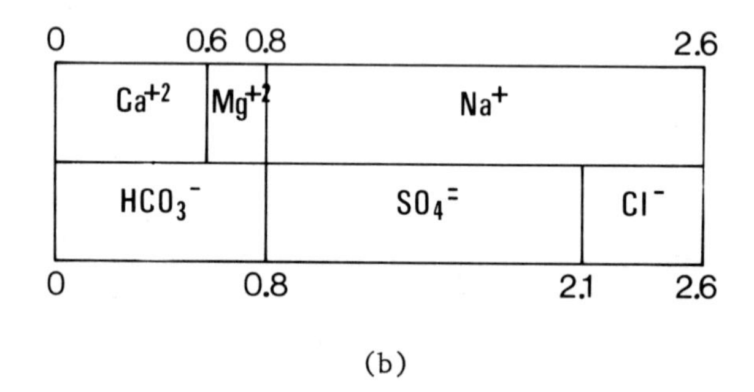

(b)

Figure 2.6. (a) Softened water before recarbonation. (b) Finished water after recarbonation.

PROBLEM 2.12

In Problem 2.11, assume that alum is added to the raw water at a concentration of 25 mg/l. Calculate the amount of lime and soda ash required for softening, assuming

35 mg/l excess lime is still required.

Solution

One mg/l alum will neutralize 0.45 mg/l bicarbonate (as $CaCO_3$) and increase the sulfate concentration (as $CaCO_3$) by an equal amount.

The bicarbonate level will therefore decrease to 263.7 mg/l as $CaCO_3$ or 5.3 meq/l. The sulfate concentration will increase to approximately 78.3 mg/l or 1.5 meq/l.

Calculate lime required.

$$Lime = meq/l\ [CO_2 + Ca(HCO_3)_2 + Mg(HCO_3)_2 + MgSO_4]$$

$$= 0.7 + 4.0 + 1.3 + 1.2 = 7.2\ meq/l$$

Lime dosage = 237 mg/l = 1976 lbs/million gallons

Calculate soda ash required.

$$meq/l\ Na_2CO_3 = 1.2$$

$$Na_2CO_3 = 1.2 \times 53 = 63.6\ mg/l$$

$$= 530\ lbs/million\ gallons$$

Alum = 25 mg/l

CaO = 237 mg/l

Na_2CO_3 = 64 mg/l

PROBLEM 2.13

Water with an analysis as shown in Table 2.3 is to be treated by a hot lime and soda ash softening process. With an excess soda ash dosage of 40 mg/l and an excess lime dosage of 5 mg/l, calcium hardness can be reduced to approximately 10 mg/l as $CaCO_3$ and $Mg(OH)_2$ to approximately 5 mg/l as $CaCO_3$. Calculate the quantity of lime and soda ash required.

Solution

No lime is required for CO_2 since the gas is evolved at the higher temperature.

Lime = meq/l $[Ca(HCO_3)_2 + Mg(HCO_3)_2 + MgSO_4]$

= 4.0 + 1.5 + 1.0 = 6.5 meq/l

Using 5 mg/l excess lime dosage,

Lime = (6.5 x 28) + 5 = 187 mg/l

= 1560 lbs/million gallons

Soda ash = 1.0 meq/l + 40 mg/l excess

= (1.0 x 53) + 40 = 93 mg/l

= 776 lbs/million gallons

PROBLEM 2.14

A water has the following analysis: pH = 7.3, alkalinity = 80 mg/l as $CaCO_3$, calcium = 35 mg/l as $CaCO_3$. Determine if the water is likely to be corrosive to an iron pipe distribution system.

Solution

Equations for carbonate system necessary to solve problem can be written as:

$$(1)\quad Ca^{+2} + CO_3^{-2} = CaCO_3 \qquad K_s = [Ca^{+2}][CO_3^{-2}] = 4.57 \times 10^{-9}$$

$$(2)\quad H^+ + HCO_3^- = H_2CO_3 \qquad K_1 = \frac{[H^+][HCO_3^-]}{[H_2CO_3]} = 4.5 \times 10^{-7}$$

$$(3)\quad H^+ + CO_3^{-2} = HCO_3 \qquad K_2 = \frac{[H^+][CO_3^{-2}]}{[HCO_3^-]} = 4.5 \times 10^{-11}$$

where K_s = solubility product constant, 25° C

K_1, K_2 = ionization constants, 25° C

Concentrations are expressed in moles/liter.

If the product of Ca^{+2} and CO_3^{-2} concentrations exceed the solubility product, then $CaCO_3$ will be precipitated out as shown by equation (1) and a protective $CaCO_3$ film will be deposited on the pipes to prevent corrosion. If the solubility product is calculated to be less than 4.57×10^{-9}, then no $CaCO_3$ layer will form and a corrosion potential exists.

The equations for alkalinity and acidity may be written as:

$$\text{Alkalinity} = HCO_3^- + 2CO_3^{-2} + OH^- - H^+$$

$$\text{Acidity} = HCO_3^- + 2H_2CO_3 - OH^- + H^+$$

Since $[CO_2]$ is essentially equal to $[H_2CO_3]$,

$$\text{Acidity} = HCO_3^- + CO_2 + H_2CO_3 - OH^- + H^+$$

In order to calculate the solubility product the $CaCO_3$ value must be determined. Equation (3) is substituted into the alkalinity equation and solved for $CaCO_3$.

$$\text{Alkalinity} = HCO_3^- + 2CO_3^{-2} + OH^- - H^+$$

$$= \frac{[H^+][CO_3^{-2}]}{K_2} + 2CO_3^{-2} + OH^- - H^+$$

$$\text{Alkalinity} = CO_3^{-2}(H^+/K_2 + 2) + OH^- - H^+$$

$$CO_3 = \frac{\text{Alkalinity} + H^+ - OH^-}{(H^+/K_2) + 2}$$

Calculate concentrations as moles/liter.

$$\text{mg/l } Ca^{+2} = 35 \text{ mg/l } CaCO_3 \times \frac{20}{50} = 14 \text{ mg/l}$$

$$[Ca^{+2}] = \frac{0.014 \text{ grams/liter}}{40.1 \text{ grams/mole}} = 3.49 \times 10^{-4} \text{ moles/liter}$$

mg/l alkalinity ($CaCO_3$) = 80 mg/l

$$[CaCO_3] = \frac{0.08 \text{ grams/liter}}{50 \text{ grams/mole}} = 1.6 \times 10^{-3} \text{ moles/liter}$$

$$pH = -\log_{10} [H^+] = 7.3$$

$$[H^+] = 5.05 \times 10^{-8} \text{ moles/liter}$$

$$pOH = 14 - 7.3 = 6.7 = -\log_{10} [OH^-]$$

$$[OH^-] = 2.0 \times 10^{-7} \text{ moles/liter}$$

Calculate CO_3 concentration.

$$CO_3 = \frac{\text{Alkalinity} + H^+ - OH^-}{(H^+/K_2) + 2}$$

$$CO_3 = \frac{1.6 \times 10^{-3} + 5.05 \times 10^{-8} - 2.0 \times 10^{-7}}{(5.05 \times 10^{-8}/5.6 \times 10^{-11}) + 2}$$

$$CO_3 = 1.78 \times 10^{-6} \text{ moles/liter}$$

Calculate solubility product.

$$[Ca^{+2}][CO_3^{-2}] = (3.49 \times 10^{-4})(1.78 \times 10^{-6})$$

$$= 0.62 \times 10^{-9}$$

Since solubility product is less than 4.57×10^{-9}, $CaCO_3$ is not likely to precipitate and the water will tend to be corrosive.

PROBLEM 2.15

Consider the two water analyses in Table 2.4. Using the Caldwell-Lawrence diagrams, determine if each water is oversaturated or undersaturated with respect to calcium carbonate. Calcium and alkalinity concentrations are given in mg/l as $CaCO_3$.

Table 2.4

WATER ANALYSES FOR PROBLEM 2.15

A	B
Calcium = 90 mg/l	Calcium = 25 mg/l
Alkalinity = 90 mg/l	Alkalinity = 40 mg/l
pH = 8.4	pH = 8.6
Temperature = 16° C	Temperature = 16° C
Total Dissolved Solids = 110 mg/l	Total Dissolved Solids = 95 mg/l

Solution

The Caldwell-Lawrence diagrams are useful for solving water softening problems and for corrosion control calculations involving calcium carbonate equilibria. Caldwell-Lawrence diagrams for 2°, 5°, 15°, and 20° C at total dissolved solids concentrations of 40, 400, and 1200 mg/l are included at the end of this chapter. These diagrams are reproduced with permission of Brown and Caldwell Consulting Engineers, Walnut Creek, California. These are modified Caldwell-Lawrence diagrams, using axes developed by R.E. Loewenthal and G.v.R. Marais of the University of Cape Town, South Africa.[3] A detailed discussion of the use of these diagrams for water softening and corrosion control calculations is found in _Water Treatment Plant Design for the Practicing Engineer_[4] edited by R.L. Sanks. A discussion of the use of these diagrams for corrosion control applications is found in _Corrosion Control By Deposition of Calcium Carbonate Films_[5] published by the American Water Works Association. Full scale diagrams (14 x 18 inches) may be ordered directly from the American Water Works Association.

The water condition is determined from the diagrams by locating the pH, alkalinity, and calcium lines as shown in Figure 2.7 for the water analyses given in Table 2.4. The 15° C and 40 mg/l TDS diagram is used as most representative of the given conditions. Note that the lines form an envelope, indicating the water is either oversaturated or undersaturated with respect to calcium carbonate. Lines that intersect at a point are exactly saturated. If the

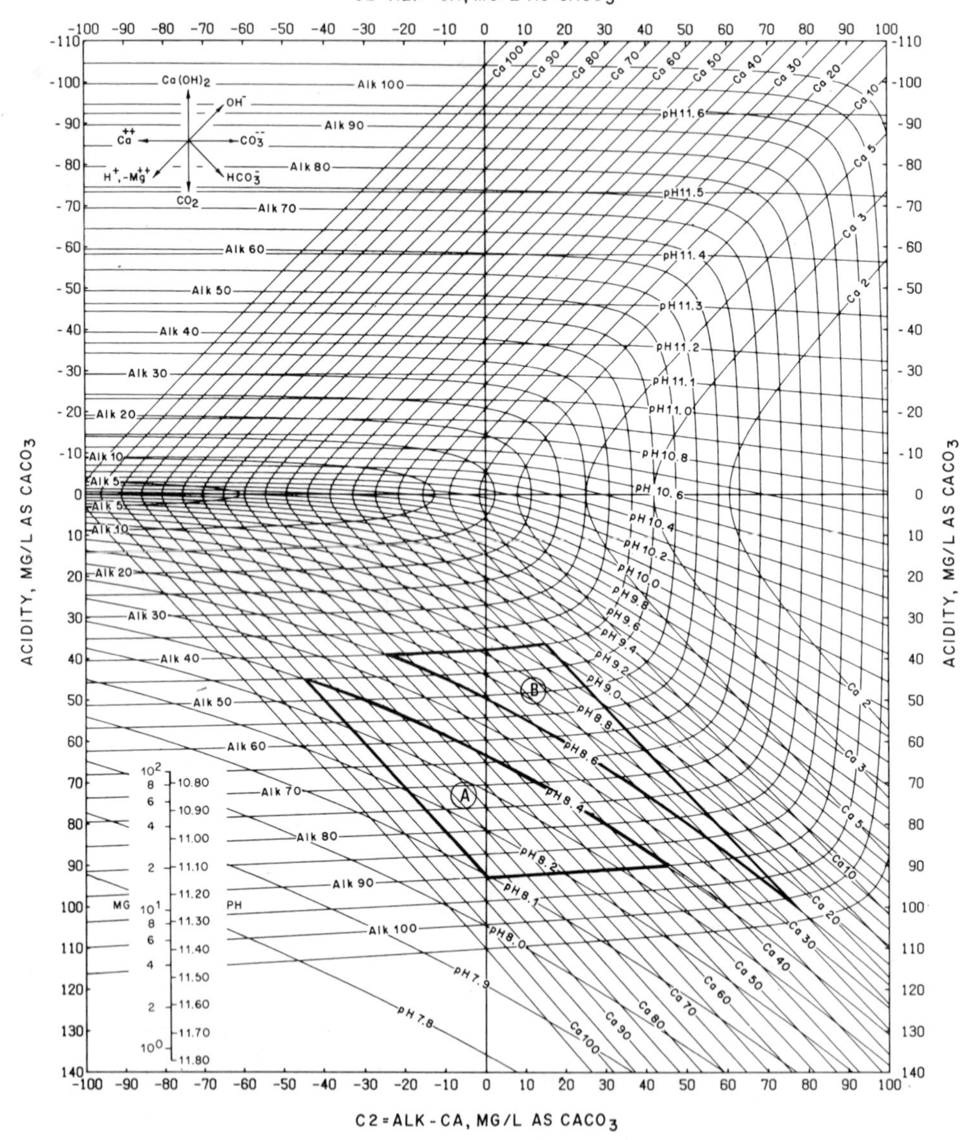

Figure 2.7. Water conditioning diagram for problem 2.15.

Caldwell-Lawrence Water Conditioning Diagram.
(Reprinted with permission of Brown and Caldwell Consulting Engineers.)

calcium value from the water analysis is greater than the calcium value found at the intersection of the pH and alkalinity lines on the diagram, the water is oversaturated. Calcium carbonate would be expected to be precipitated. If the calcium value is less than that determined from the diagram, the water is undersaturated.

Water A is oversaturated since the calcium value read from the Caldwell-Lawrence diagram is 45 mg/l, less than the 90 mg/l from the water analysis. Water B is undersaturated. The water analysis gives a calcium value of 25 mg/l and a calcium value of 65 mg/l is read from the diagram.

The diagrams could also be used for the water given in Problem 2.14. In this case the lines can not be located on the diagram (25° C, 40 mg/l TDS) and would be found somewhere to the left of the diagram. However, it becomes obvious that the calcium value at the intersection of 80 mg/l alkalinity and 7.3 pH would be in excess of 100 mg/l. Since the water analysis shows a calcium value of only 35 mg/l, the water would be undersaturated and calcium carbonate would not be expected to precipitate under these conditions.

In using the diagrams the following guidelines are helpful:

(1) Concentrations for calcium and alkalinity are mg/l as $CaCO_3$.

(2) The acidity of a water is found at the intersection of the alkalinity and pH lines.

(3) The C2 value is found at the intersection of the calcium and alkalinity lines.

(4) Alkalinity = $HCO_3^- + CO_3^{-2} + OH^- - H^+$

(5) Acidity = $CO_2 + H_2CO_3 + HCO_3^- + H^+ - OH^-$

PROBLEM 2.16

For the two water analyses given in Problem 2.15, determine the saturation point of the water with respect to calcium carbonate and the maximum theoretical amount of calcium carbonate that will be precipitated.

Solution

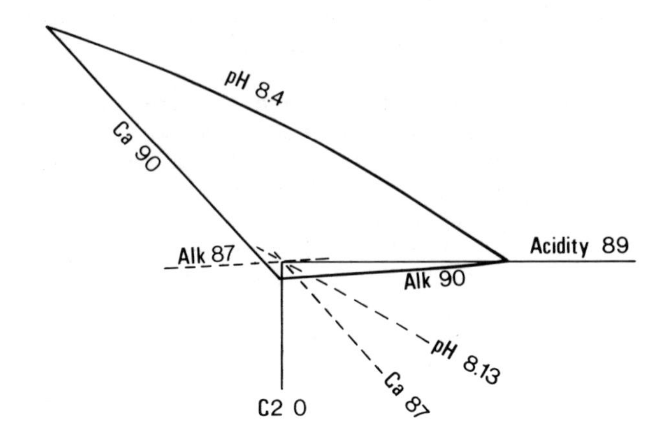

Water A

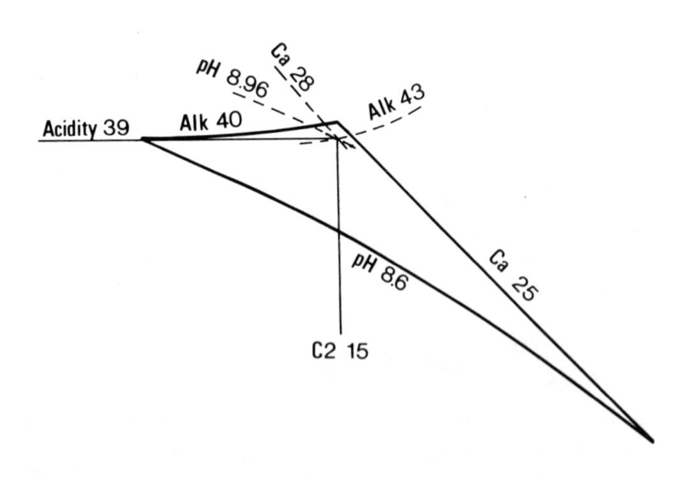

Water B

Figure 2.8. Determination of water condition. Problem 2.16.

The solution is shown in Figure 2.8. The condition of the saturated water is determined by the intersection of the acidity and C2 lines. The acidity line is the horizontal line located through the intersection of the alkalinity and pH lines. The C2 line is the vertical line located through the intersection of the calcium and alkalinity lines.

For water A, the saturated state determined from the Caldwell-Lawrence diagram is pH = 8.13, calcium = 87, alkalinity = 87. For water B, pH = 8.96, calcium = 28, alkalinity = 43.

The theoretical amount of calcium carbonate that can be precipitated is the difference between the calcium value of the water analysis and the calcium value at the saturation point. Water B is undersaturated and no calcium carbonate would be expected to precipitate. Water A is oversaturated and the maximum amount of calcium carbonate precipitated would be calculated as,

$$90 \text{ mg/l} - 87 \text{ mg/l} = 3 \text{ mg/l as } CaCO_3$$

PROBLEM 2.17

A water has the following analysis: alkalinity = 45 mg/l, calcium = 35 mg/l, pH = 8.4. Use the Caldwell-Lawrence diagram for 15° C, 40 mg/l TDS to determine the final water composition after addition of the following chemicals. Calculate the theoretical amount of calcium carbonate precipitated. Chemical dosage is mg/l as $CaCO_3$.

(a) 10 mg/l $Ca(OH)_2$

(b) 10 mg/l $Ca(OH)_2$ and 10 mg/l Na_2CO_3

Solution

(a) The water is undersaturated since the calcium value at the intersection of the alkalinity and pH lines (90 mg/l) on the Caldwell-Lawrence diagram is greater than the measured calcium value (35 mg/l). Change in alkalinity and acidity is defined as,

$$\text{Alkalinity} = HCO_3^- + CO_3^{-2} + OH^- - H^+$$

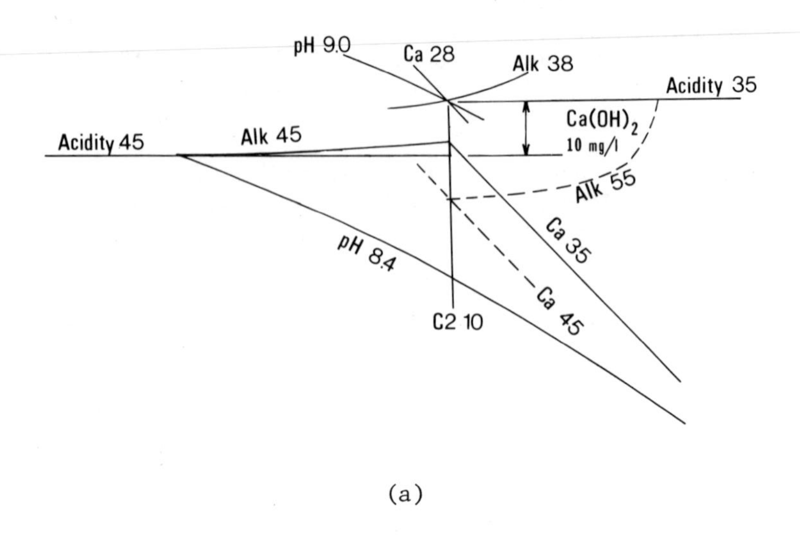

(a)

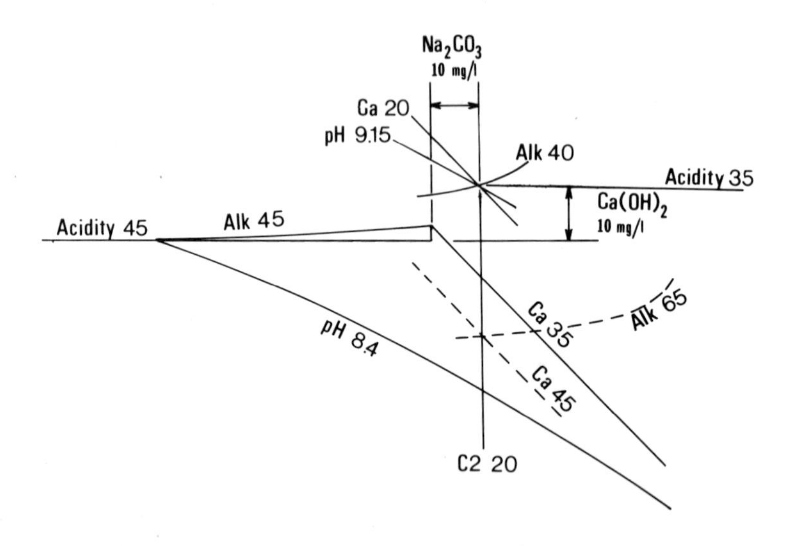

(b)

Figure 2.9. (a) Water condition for addition of 10 mg/l $Ca(OH)_2$ (b) Water condition for addition of 10 mg/l $Ca(OH)_2$ and 10 mg/l Na_2CO_3.

$$\text{Acidity} = CO_2 + H_2CO_3 + HCO_3^- + H^+ - OH^-$$

Addition of 10 mg/l calcium hydroxide increases alkalinity by 10 mg/l, increasing calcium by 10 mg/l, and decreases acidity by 10 mg/l. Initial acidity is determined from the Caldwell-Lawrence diagram as 45 mg/l.

Alkalinity = 45 mg/l + 10 mg/l = 55 mg/l as $CaCO_3$

Acidity = 45 mg/l - 10 mg/l = 35 mg/l as $CaCO_3$

Calcium = 35 mg/l + 10 mg/l = 45 mg/l as $CaCO_3$

Using these calculated values for alkalinity, acidity, and calcium, the saturation state of the water after chemical addition is determined as the intersection of the C2 and acidity lines as shown in Figure 2.9(a). At saturation, pH = 9.0, Calcium = 28 mg/l, Alkalinity = 38 mg/l.

The theoretical amount of calcium carbonate that can be precipitated is the difference between the calcium value of the water before chemical addition (35 mg/l) and the calcium value at saturation (28 mg/l), or 7 mg/l as $CaCO_3$.

(b) For addition of 10 mg/l $Ca(OH)_2$ and 10 mg/l Na_2CO_3

Alkalinity = 45 mg/l + 10 mg/l + 10 mg/l = 65 mg/l as $CaCO_3$

Acidity = 45 mg/l - 10 mg/l = 35 mg/l as $CaCO_3$

Calcium = 35 mg/l + 10 mg/l = 45 mg/l as $CaCO_3$

The saturated state of the water at the intersection of the C2 and acidity is determined from Figure 2.9(b) as pH = 9.15, Calcium = 20 mg/l, Alkalinity = 40 mg/l.

$CaCO_3$ precipitated = 35 mg/l - 20 mg/l = 15 mg/l as $CaCO_3$

The required amount of calcium carbonate precipitated to prevent corrosion will vary with the conditions of each distribution system. In general, the following treated water conditions are recommended:[6]

1. The theoretical amount of calcium carbonate precipitated should be in the range of 4-10 mg/l as $CaCO_3$.

2. Calcium and alkalinity values should each equal at least 40 mg/l as $CaCO_3$.

3. The pH should ideally be in the range of 6.8-7.3, and pH in the range of 8.0-8.5 should be avoided. The pH is critical to prevent $Mg(OH)_2$ scaling in domestic water heaters.

4. The ratio of alkalinity/(Cl^- + SO_4^{-2}) should be at least 5:1 expressed as mg/l $CaCO_3$.

5. Adequate water velocity should be provided.

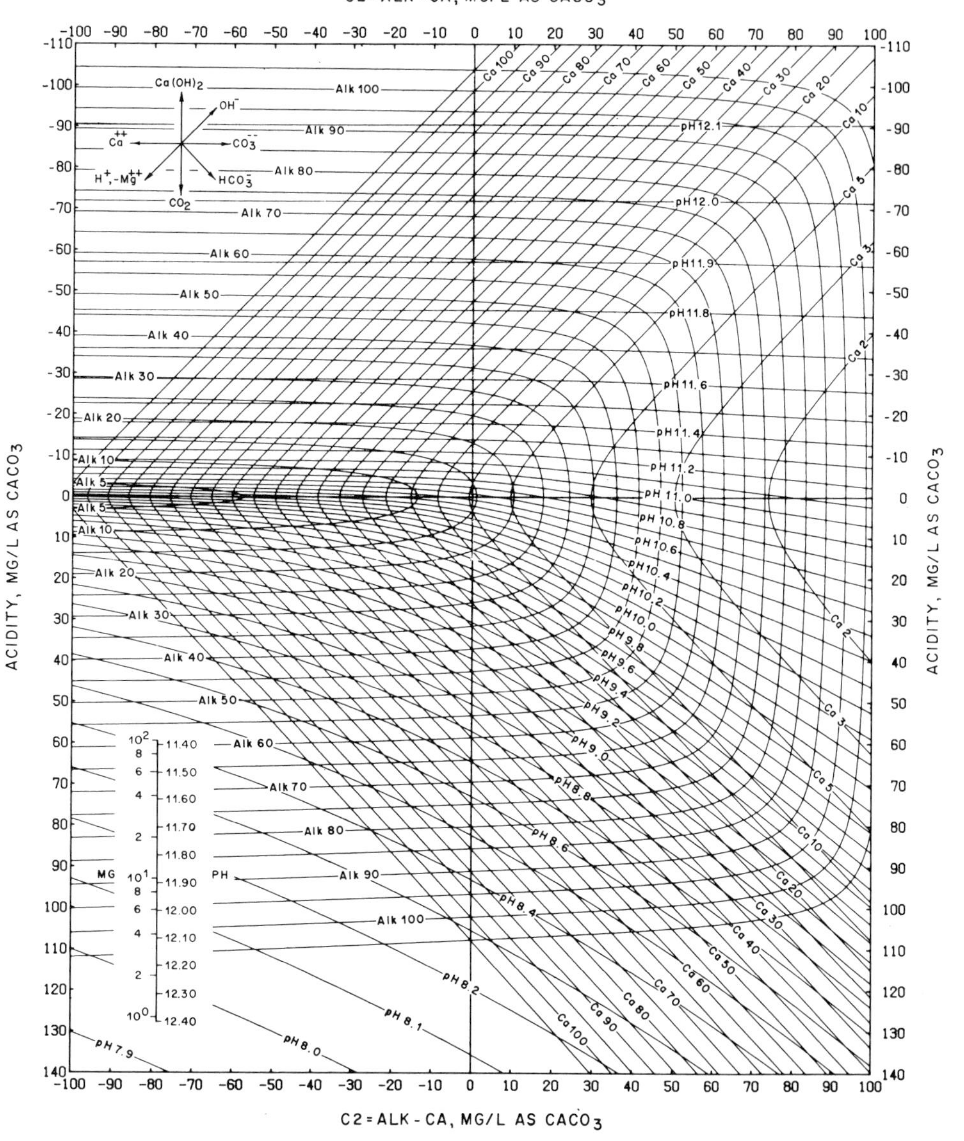

Caldwell-Lawrence Water Conditioning Diagram.
(Reprinted with permission of Brown and Caldwell Consulting Engineers.)

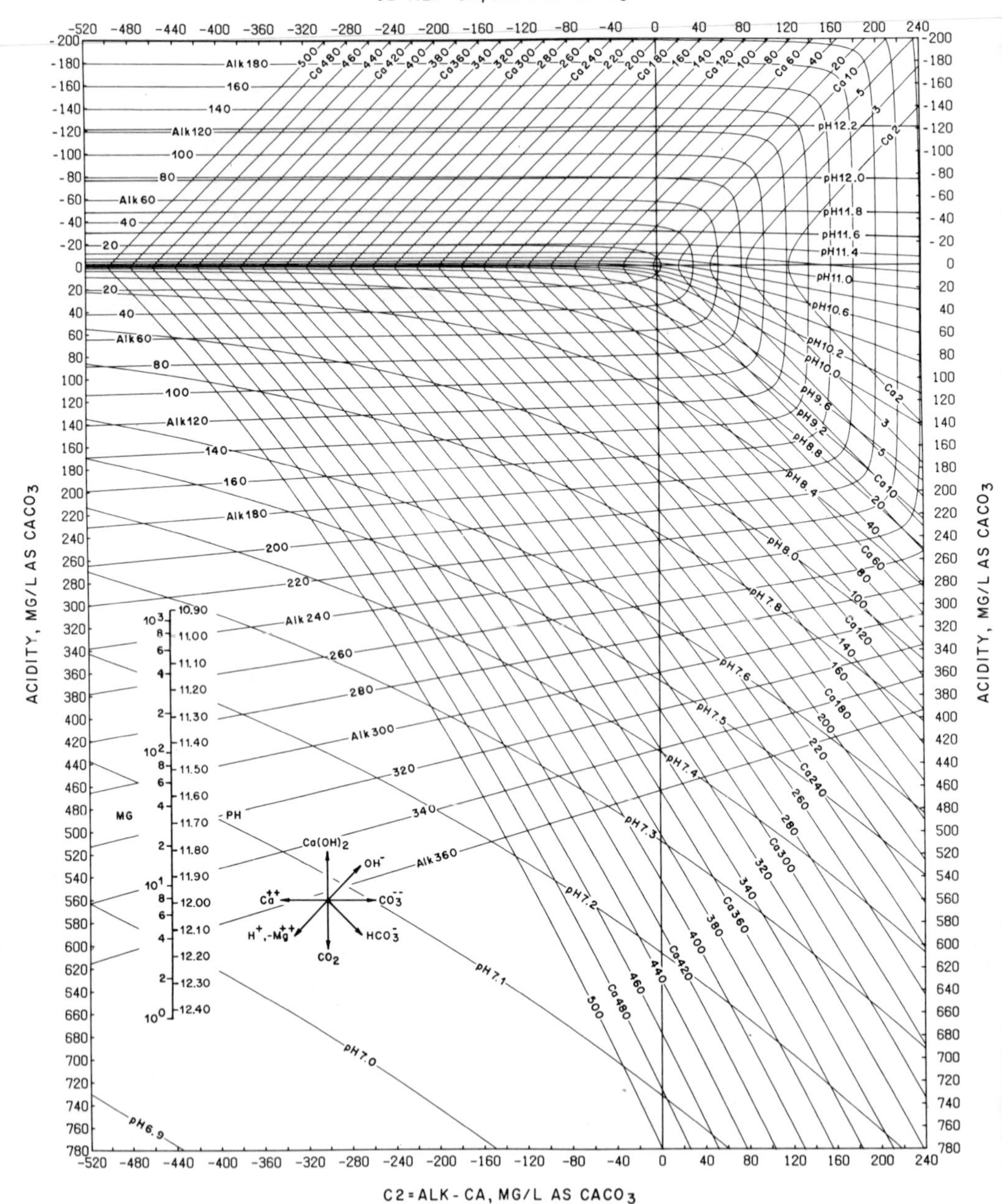

Caldwell-Lawrence Water Conditioning Diagram.
(Reprinted with permission of Brown and Caldwell Consulting Engineers.)

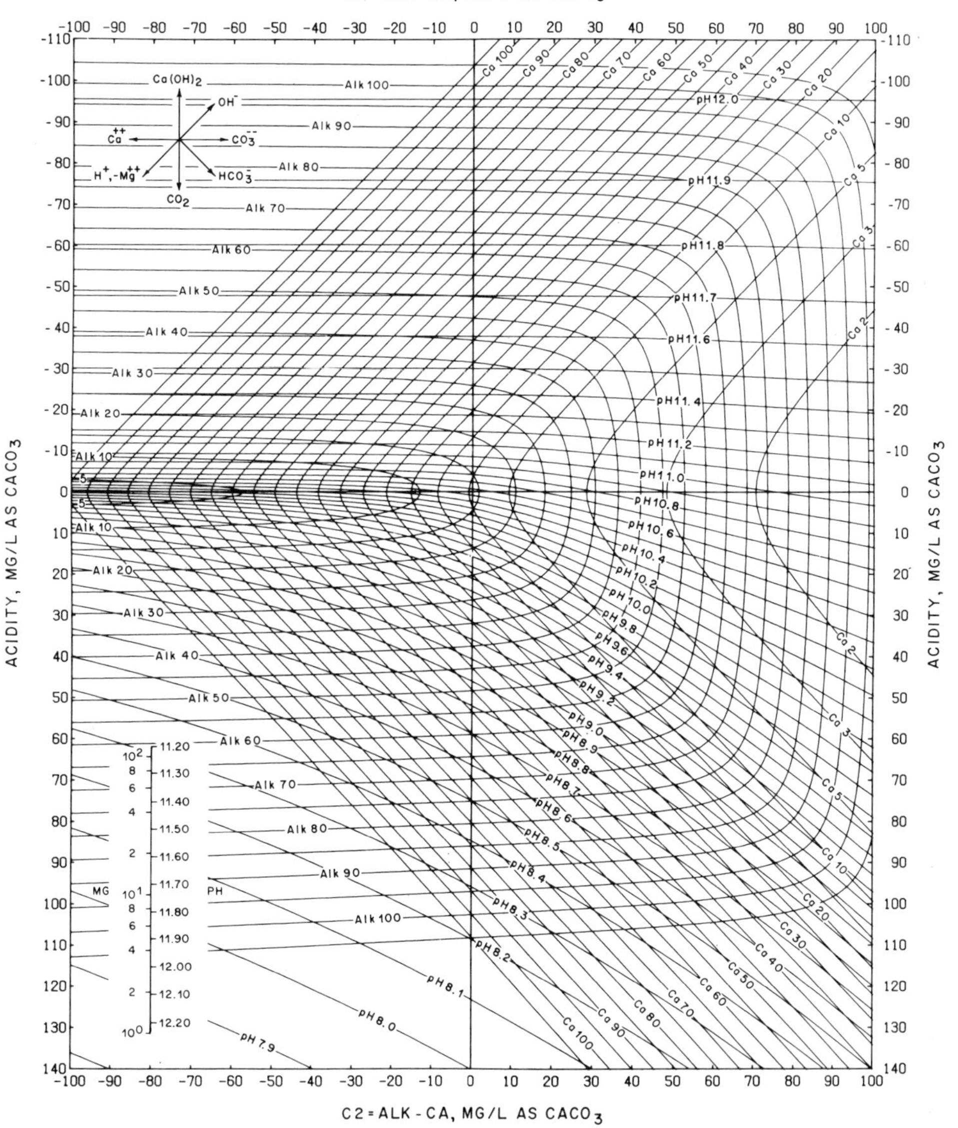

Caldwell-Lawrence Water Conditioning Diagram.
(Reprinted with permission of Brown and Caldwell Consulting Engineers.)

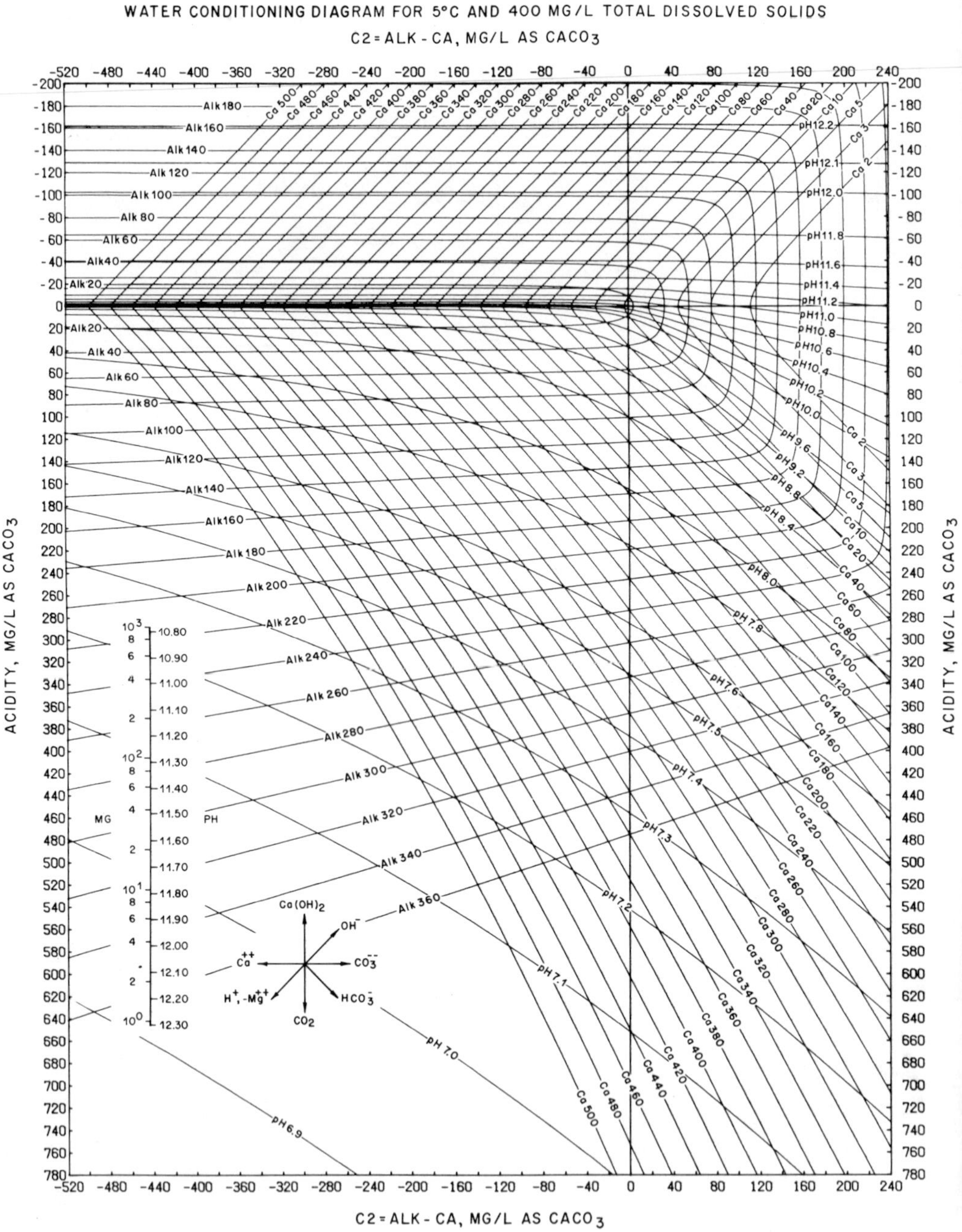
WATER CONDITIONING DIAGRAM FOR 5°C AND 400 MG/L TOTAL DISSOLVED SOLIDS
C2 = ALK - CA, MG/L AS CACO3
ACIDITY, MG/L AS CACO3
C2 = ALK - CA, MG/L AS CACO3
MG
PH
Ca(OH)2
OH⁻
Ca⁺⁺
CO3⁻⁻
H⁺, -Mg⁺⁺
HCO3⁻
CO2

Caldwell-Lawrence Water Conditioning Diagram.
(Reprinted with permission of Brown and Caldwell Consulting Engineers.)

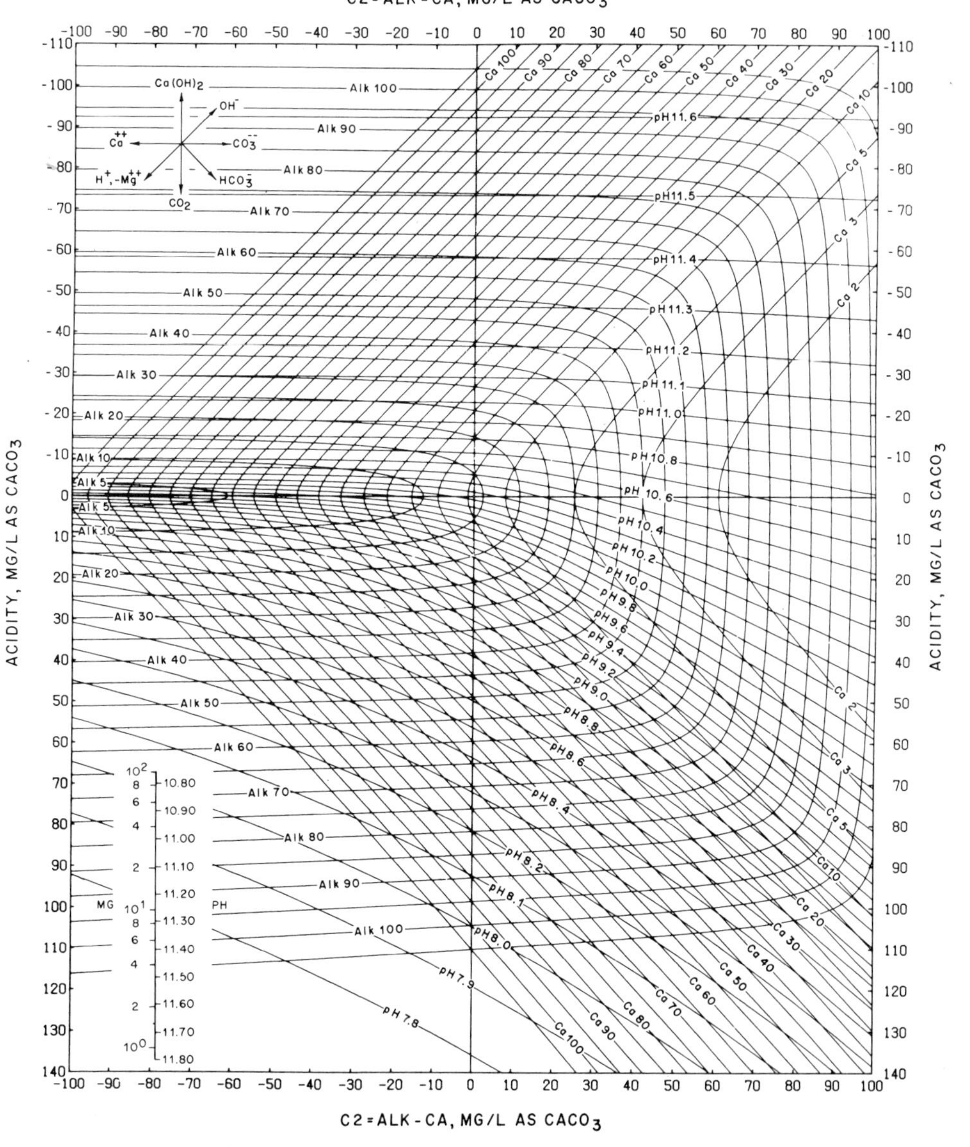

Caldwell-Lawrence Water Conditioning Diagram.
(Reprinted with permission of Brown and Caldwell Consulting Engineers.)

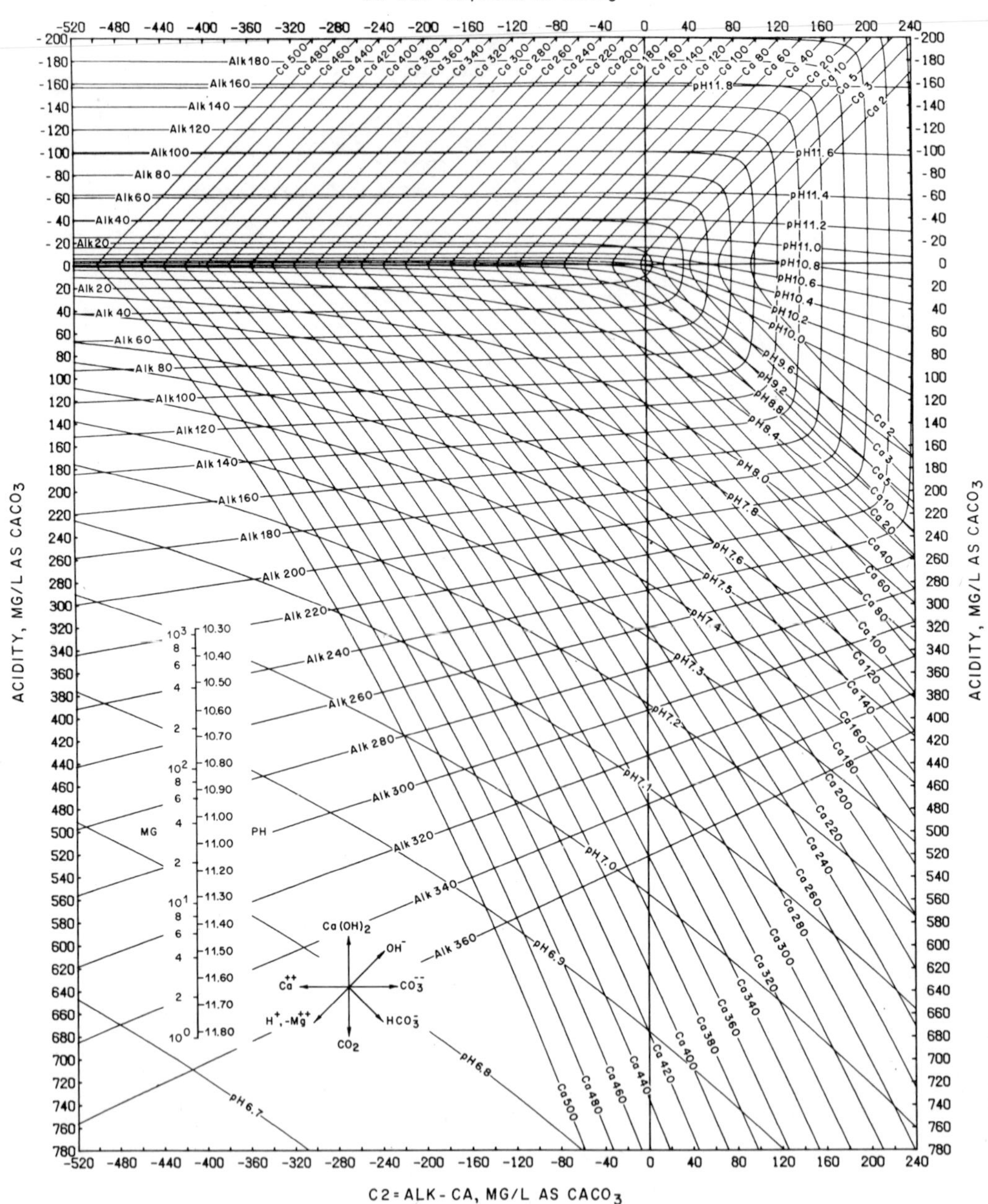

Caldwell-Lawrence Water Conditioning Diagram.
(Reprinted with permission of Brown and Caldwell Consulting Engineers.)

WATER CONDITIONING DIAGRAM FOR 15°C AND 1200 MG/L TOTAL DISSOLVED SOLIDS

C2 = ALK - CA, MG/L AS CACO$_3$

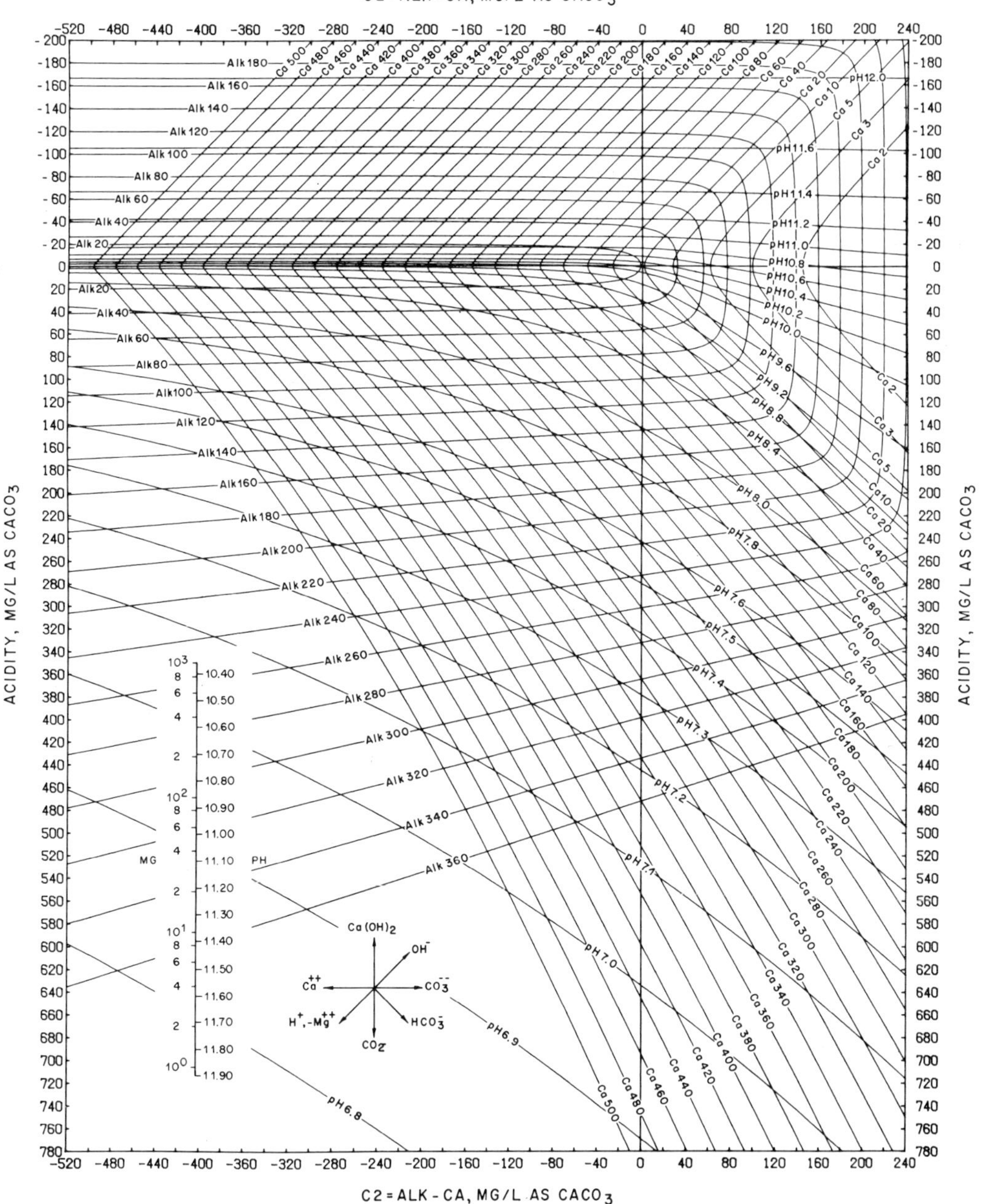

Caldwell-Lawrence Water Conditioning Diagram.
(Reprinted with permission of Brown and Caldwell Consulting Engineers.)

WATER CONDITIONING DIAGRAM FOR 25°C AND 40 MG/L TOTAL DISSOLVED SOLIDS

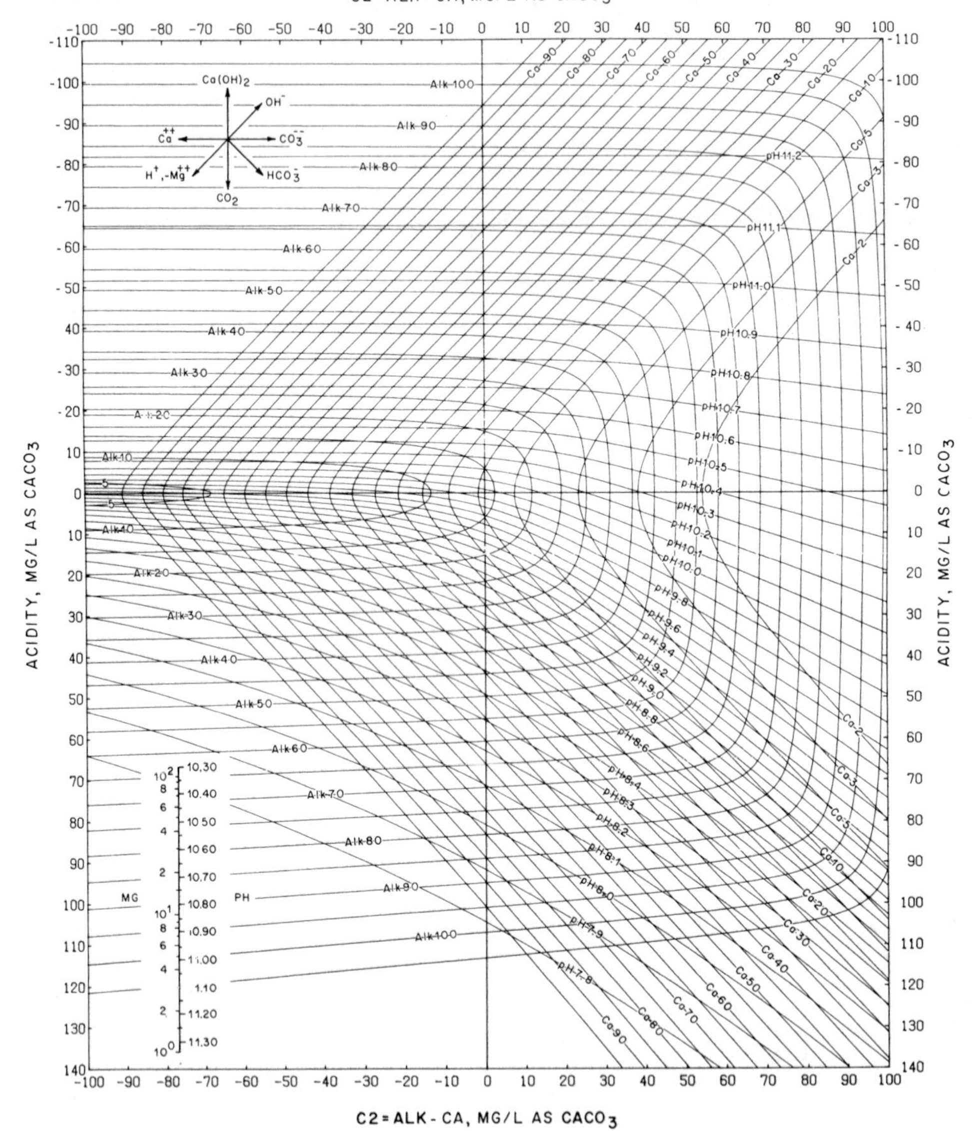

Caldwell-Lawrence Water Conditioning Diagram.
(Reprinted with permission of Brown and Caldwell Consulting Engineers.)

WATER CONDITIONING DIAGRAM FOR 25°C AND 400 MG/L TOTAL DISSOLVED SOLIDS

C2 = ALK - CA, MG/L AS CACO3

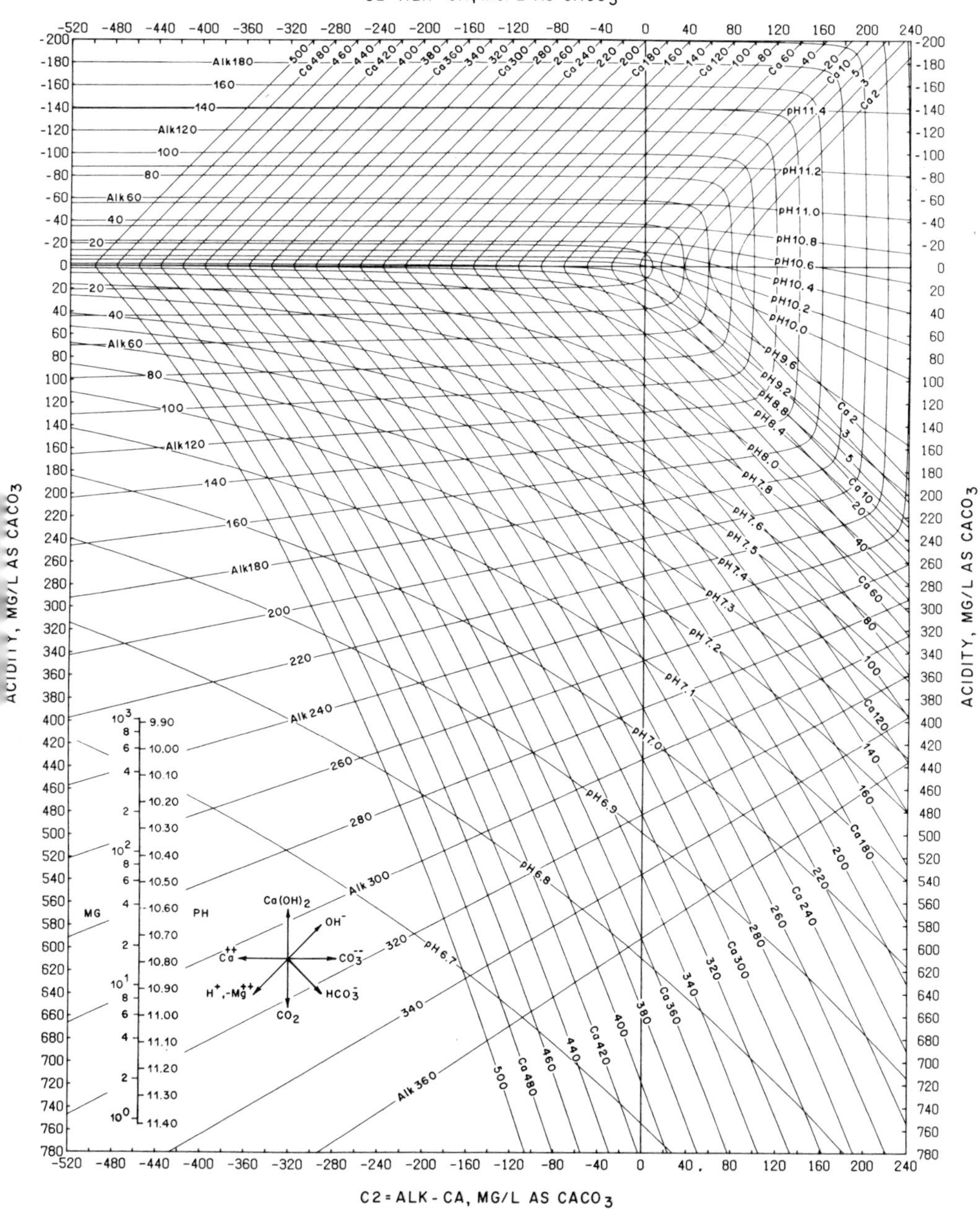

Caldwell-Lawrence Water Conditioning Diagram. (Reprinted with permission of Brown and Caldwell Consulting Engineers.)

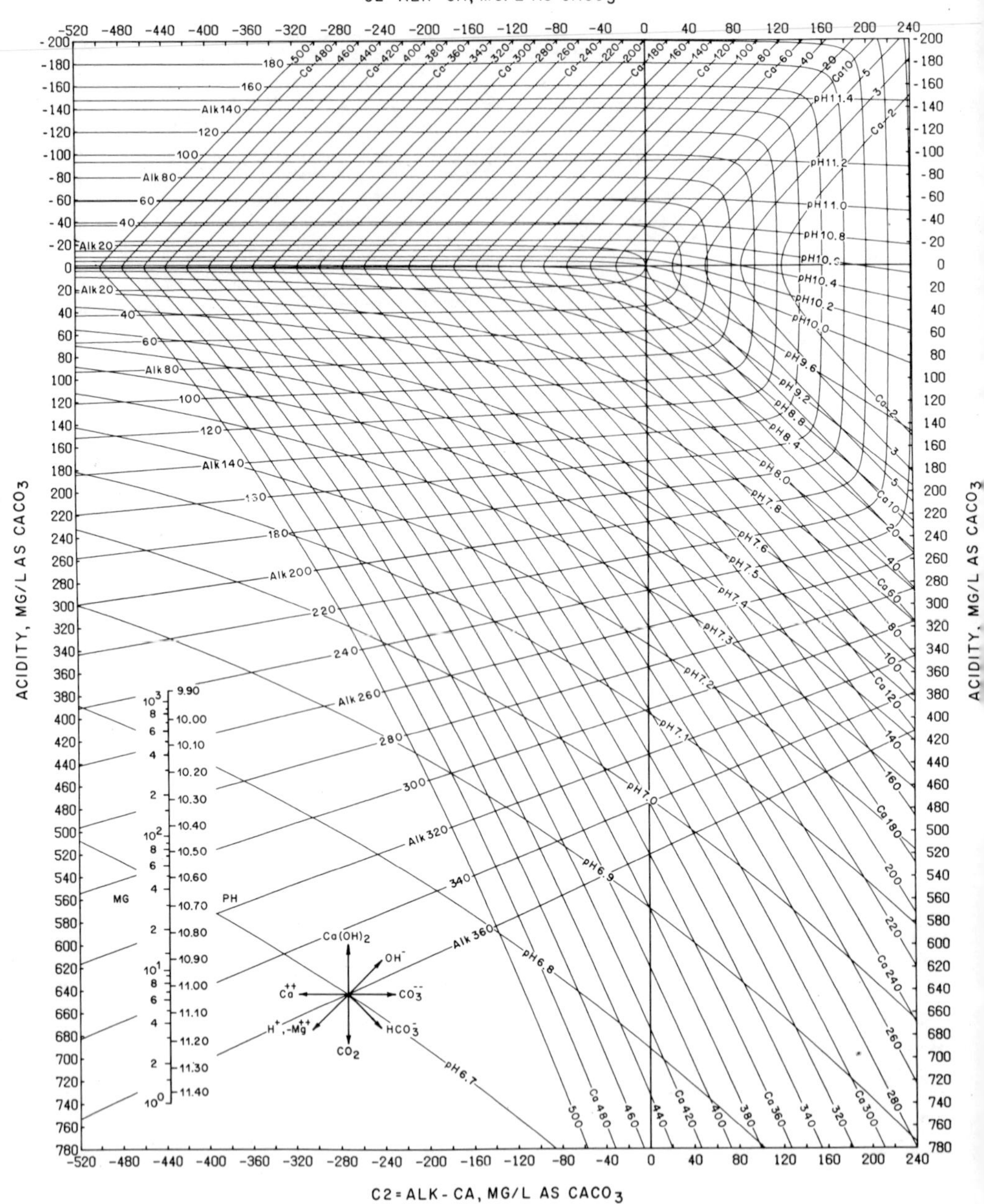

Caldwell-Lawrence Water Conditioning Diagram.
(Reprinted with permission of Brown and Caldwell Consulting Engineers.)

References

1. Fair, G.M. and L.P. Hatch, "Fundamental Factors Governing the Streamline Flow of Water through Sand," J. Am. Water Works Assoc., 25:1551, 1933.

2. Thomas R. Camp, "Water Treatment" in Handbook of Applied Hydraulics, Calvin Victor Davis, Editor-in-Chief, McGraw-Hill Book Company, Inc., New York, 1942.

3. Loewenthal, R.E. and G.v.R. Marais, The Carbonic System in Water Treatment, Research Report No. W4, University of Cape Town, Rondebosch, Cape Town, South Africa, 1973.

4. Sanks, Robert L., Water Treatment Plant Design for the Practicing Engineer, Ann Arbor Science Publishers, Inc., Ann Arbor, Michigan, 1978.

5. Merrill, Douglas T., and Robert L. Sanks, Corrosion Control By Deposition of $CaCO_3$ Films, American Water Works Association, Denver, Colorado, 1978.

6. Ibid., pp. 4-5, 37-38.

Additional References

1. Hardenbergh, W.A. and Edward R. Rodie, Water Supply and Waste Disposal, International Textbook Company, Scranton, Pennsylvania, 1961.

2. Recommended Standards for Water Works, published by Health Education Service, Inc., Albany, New York, 1976.

3. Clark, John W., Warren Viessman, Jr., Mark J. Hammer, Water Supply and Pollution Control, Harper and Row Publishers, Inc., New York, 1977.

4. Sawyer, Clair N. and Perry L. McCarty, Chemistry for Sanitary Engineers, McGraw-Hill Book Company, New York, New York, 1967.

5. "Form and Procedures for Fire Flow Tests - Committee Report," Journal American Water Works Association, May, 1976, pp. 264-268.

Chapter 3

WASTEWATER TREATMENT

PROBLEM 3.1

The following variations of the activated sludge process are commonly used in wastewater treatment:

Conventional Activated Sludge

High Rate

Complete Mix

Step Aeration

Extended Aeration

Contact Stabilization

Pure Oxygen

Identify the following design parameters for each process applicable to the treatment of domestic wastewater: percent BOD_5 removal; aeration detention time; F/M ratio; aerator loading; pounds sludge produced/pound BOD_5 removed; oxygen demand (pounds oxygen/pound BOD_5 applied).

Solution

Design parameters are summarized in Table 3.1. Values may vary with different operating conditions.

Table 3.1

DESIGN PARAMETERS FOR ACTIVATED SLUDGE PROCESS

Process Modification	BOD_5 Removal	Aeration Detention Time (Hours)	F/M Ratio lbs BOD_5/lb MLSS	Aerator Loading lbs BOD_5/1000 ft^3	lbs sludge/lb BOD_5 removed	lbs O_2/lb BOD_5 applied
Conventional	>85%	6-8	0.2-0.4	20-40	0.4-0.55	1.0-1.1
High Rate	75-85%	0.5-2.5	0.5-1.5	100-400 +	0.75-1.0	0.4-0.5
Step Aeration	>85%	3-5	0.2-0.4	40-60	0.55	1.0-1.1
Contact Stabilization	>80%	1.5-3.0	0.2-0.6	50-75	0.55	1.0-1.1
Extended Aeration	85-95%	18-24	0.05-0.15	12.5-25	0.10-0.15	1.5-1.8
Complete Mix	85-95%	4-6	0.2-0.6	50-120	0.55-0.65	1.0-1.1
Pure Oxygen	85-95%	2-4	0.5-0.9	100-250	0.45-0.55	1.3-1.8

PROBLEM 3.2

A treatment plant is designed using a conventional activated sludge process. Design population is 29,500 and industrial flow is 250,000 gallons per day. Stream standards require an effluent BOD_5 and total suspended solids concentration of 30 mg/l. Based on an analysis of water use records adjusted for infiltration a flow rate of 100 gallons per capita per day will be used for design. Develop the design criteria for each of the following treatment units: raw sewage pumping station, grit removal, primary and final clarifiers, aeration tank, and chlorine contact tank.

Solution

Assume the following design criteria are applicable:

Domestic sewage flow = 100 gal/cap/day

Peak flow = average daily flow x 2.5

BOD loading = 0.17 lbs/cap/day

TSS loading = 0.20 lbs/cap/day

Domestic flow = 29,500 x 100 = 2,950,000 gal/day
Industrial flow = 250,000 gal/day

Design flow rate = 3.2 mgd
Peak flow rate = 8.0 mgd

Population equivalent = 32,000 persons

BOD loading = 0.17 x 32,000 = 5440 lbs/day
TSS loading = 0.20 x 32,000 = 6400 lbs/day

Raw Sewage Pumping Station

Design for peak flow rate with one pump as standby.

3.2 mgd = 2222 gpm
8.0 mgd = 5556 gpm

Use 4 pumps, 1850 gpm capacity each.

Calculate maximum wet well storage for 10 minute detention time at average daily flow rate.

$$\frac{3.2 \times 10^6 \text{ gal}}{1440 \text{ min}} \times 10 \text{ min} \times 0.134 \frac{\text{ft}^3}{\text{gal}} = 2970 \text{ ft}^3$$

Grit Removal

Use 2 units, each designed for 4.0 mgd to handle peak flow.

Design for 100 mesh grit size and specific gravity of 2.0. From WPCF Manual of Practice No. 8, Sewage Treatment Plant Design, 1959, maximum allowable overflow rate is 19,600 gpd/sq. ft.

$$\frac{4.0 \times 10^6 \text{ gpd}}{19{,}600 \text{ gpd/ft}^2} = 204 \text{ ft}^2\text{/unit}$$

Minimum velocity of 1.0 ft/sec recommended.

Primary Clarifier

Use design criteria from Recommended Standards for Sewage Works (Ten State Standards), 1973.

Weir overflow rate = 10,000 to 15,000 gallons per lineal foot

BOD removal = 32% at 1000 gpd/ft^2 settling rate

Assume 40% suspended solids removal.

$$\text{Surface area} = \frac{3.2 \times 10^6 \text{ gpd}}{1000 \text{ gpd/ft}^2} = 3200 \text{ ft}^2$$

Use two 45 foot diameter clarifiers with 10 foot side water depth (SWD).

Total surface area = 3180 ft^2

$$\text{Weir rate} = \frac{3.2 \times 10^6 \text{ gpd}}{282 \text{ ft}} = 11{,}350 \text{ gpd/lineal foot}$$

$$\text{Volume} = 3180 \text{ ft}^2 \times 10 \text{ ft} = 31{,}800 \text{ ft}^3$$

$$\text{Detention time} = \frac{31{,}800 \times 7.48 \times 24}{3.2 \times 10^6} = 1.8 \text{ hours}$$

Calculate BOD loading to aeration at 32% removal rate.

5440 lbs/day x (1 - 0.32) = 3700 lbs/day

Aeration

Use design criteria from Ten State Standards.

Detention time = 6 hours

Aerator loading = 40 lbs BOD/1000 ft^3

BOD loading to aeration = 3700 lbs/day

Calculate tank volume at loading of 40 lbs/1000 ft^3.

$$\frac{3700 \text{ lbs}}{40 \text{ lbs}} \times 1000 \text{ ft}^3 = 92{,}500 \text{ ft}^3$$

Check detention time $= \dfrac{92{,}500 \times 7.48 \times 24}{3.2 \times 10^6} = 5.2$ hours

Use tank volume of 107,000 ft^3 required for 6 hour detention time.

Check F/M ratio (lbs BOD applied/lb MLVSS under aeration). Assume F/M ratio in the range of 0.25 to 0.30 is desired for operation.

Assume MLSS concentration = 2750 mg/l and
MLVSS/MLSS ratio = 0.8

$$F/M = \frac{3700}{2750 \times 0.8 \times 8.34 \times 0.107 \times 7.48} = 0.25$$

Use 2 basins with SWD = 14 feet

Total area = 107,000 ft^3/14 ft = 7645 ft^2

Assume each basin 44 ft x 88 ft.

Total volume = 108,400 ft^3

Calculate aeration requirements.

Assume 1.1 lbs O_2 required/lb BOD applied.

O_2 = 1.1 x 3700 = 4070 lbs/day = 170 lbs/hr

Assume aerators with an oxygen transfer rate of 2.5 lbs O_2/hp-hr.

Aerator horsepower = 170/2.5 = 68 hp
Add 50% safety factor = 102 hp

Use 4 aerators, each 25 hp. Optimum size and spacing should be based on equipment manufacturer's recommendation. Check air volume supplied for 100 hp.

O_2 supplied = 100 x 2.5 x 24 = 6000 lbs/day

$$\text{Volume air} = \frac{6000 \text{ lbs/day}}{0.0724 \times 0.22 \times 0.07}$$

where 0.0724 = specific weight of air at 1000 ft elevation, lbs/cu. ft.

0.22 = weight ratio oxygen to air

0.07 = oxygen transfer efficiency (assumed)

Volume air = 5.38×10^6 ft^3

$$\frac{5.38 \times 10^6}{3700} = 1455 \text{ ft}^3\text{/lb BOD applied}$$

Ten State Standards recommend 1500 cu. ft. of air/lb BOD applied. Calculated value is considered acceptable for biological treatment requirement.

Aeration capacity should also be evaluated to ensure adequate mixing is provided. Assume that 0.5 to 1.0 hp/1000 cu. ft. of aeration tank volume is required.

$$\frac{100 \text{ hp}}{107{,}000 \text{ ft}^3} = 0.93 \text{ hp/1000 ft}^3$$

Final Clarifier

Use design criteria from Ten State Standards.

Surface loading = 800 gpd/ft^2

Weir overflow rate = 10,000 to 15,000 gallons per lineal foot

$$\frac{3.2 \times 10^6 \text{gpd}}{800 \text{ gpd/ft}^2} = 4000 \text{ ft}^2$$

Use two 51 foot diameter clarifiers with 8 foot side water depth.

Total area = 4085 ft^2

$$\text{Weir rate} = \frac{3.2 \times 10^6 \text{ gpd}}{320 \text{ ft}} = 10{,}000 \text{ gpd/lineal foot}$$

$$\text{Total volume} = 4085 \text{ ft}^2 \times 8 \text{ ft} = 32{,}680 \text{ ft}^3$$

$$\text{Detention time} = \frac{24 \times 32{,}680 \times 7.48}{3.2 \times 10^6} = 1.8 \text{ hours}$$

Check solids loading. Assume MLSS = 2750 mg/1 and return sludge flow rate = 30%

$$\frac{2750 \times 8.34 \times 4.2}{4085 \text{ ft}^2} = 24 \text{ lbs/ft}^2$$

Chlorine Contact Tank

Design for 15 minutes detention time at maximum flow.

$$\frac{8.0 \times 10^6}{1440} \times 15 \times \frac{1}{7.48} = 11{,}140 \text{ ft}^3$$

Use two tank with 8 foot side water depth.

Total area required = 11,140 ft^3/8 ft = 1395 ft^2

Assume each tank 19 ft x 37 ft.

Total area = 1400 ft^2

PROBLEM 3.3

The 5-day BOD of an industrial waste is 235 mg/1. The first stage ultimate oxygen demand is 350 mg/1.

(a) At what rate is the waste being oxidized (assume temperature = 20° C)?

At this rate 50% of the oxygen demand will be exerted in how many hours?

(b) What is the oxidation rate at 25° C?

Solution

(a) $\% \text{ BOD oxidized} = \dfrac{235 \text{ mg/l}}{350 \text{ mg/l}} = 67.1\%$

% BOD remaining = 32.9%

$$\frac{L_T}{L} = 10^{-kt}$$

where t = time in days

L = ultimate BOD

L_T = BOD remaining at time t

k = reaction rate

$$\log_{10} \frac{L_T}{L} = -kt$$

$$\log_{10} \frac{0.329 \times 350}{350} = -k(5)$$

$$k = 0.097$$

Calculate time required for 50% oxygen demand.

$$\log_{10} 0.5 = -0.097\, t$$

$$t = 3.10 \text{ days} = 74.4 \text{ hours}$$

(b) Calculate oxidation rate at 25° C.

$$k_T = k_{20}(1.047)^{T-20}$$

where T = temperature in degrees C

k_{20} = reaction rate at 20° C

k_T = reaction rate at temperature T

$$k_T = 0.097\ (1.047)^{25-20} = 0.122$$

PROBLEM 3.4

Dilution water containing 8.0 mg/l dissolved oxygen was used for a BOD determination. At the end of the 5 days incubation period 3 mg/l dissolved oxygen had been depleted in the bottle. A 2% dilution was used.

(a) Calculate the BOD.

(b) From the result of (a), calculate the 2-day and the ultimate BOD.

(c) Why are samples incubated at 20° in the dark for 5 days?

(d) Define first stage BOD.

(e) What is the effect of nitrification in the BOD test?

(f) Identify two test conditions that may affect the rate constant "k".

Solution

(a) 5-day BOD = 3 mg/l/0.02 = 150 mg/l

(b) Calculate ultimate BOD.

$$y = L\left[1 - e^{-kt}\right]$$

where y = BOD exerted at time t

L = ultimate BOD

Assume k = 0.23

$$150 \text{ mg/l} = L\left[1 - e^{-.23(5)}\right] = L(1 - 0.316) = 219 \text{ mg/l}$$

Calculate 2-day BOD.

$$y = 219\left[1 - e^{-.23(2)}\right] = 219\ (1 - 0.63) = 81 \text{ mg/l}$$

(c) Complete stabilization may require a period too long for practical purposes. Incubation for five days at 20° C is a recognized standard. The 20° approximates an average temperature for slow moving streams in summer.

All light should be excluded during incubation to prevent oxygen formation by algae in the sample.

(d) First stage BOD is the carbonaceous oxygen demand of the waste.

(e) Nitrification is the sequential nitrogenous oxidation of wastewater which exerts a second stage of oxygen demand in addition to the first stage carbonaceous oxygen demand. It may require five to ten days to establish a nitrifying bacteria population and exert a measureable oxygen demand.

(f) The rate constant "k" may vary significantly with the type of waste and also with temperature.

PROBLEM 3.5

You are requested to prepare a preliminary design for a wastewater treatment system for Pine Run State Park. The system will serve 15 cabins with an average of 4 persons per cabin. Water quality standards require an effluent with a 30-day average of 30 mg/l for BOD_5 and total suspended solids. Assume that an on-lot sewage system is not feasible.

(a) What type of treatment system would you recommend? Justify your conclusions.

(b) Calculate treatment plant design flow rate. Show all calculations.

(c) Given the above information, what size collection sewers might reasonably be required? Show all calculations.

Solution

(a) Installation of a factory built or "package" type treatment plant using the extended aeration process is recommended. This offers the advantages of simplicity and ease of operation, low sludge yield, and good effluent quality. Since flow rates and organic loading may be quite variable, standby aeration capacity is recommended for peak periods. Disinfection prior to discharge is also recommended. A lagoon may also be selected as a suitable treatment method due to simplicity and ease of operation.

(b) Design for 15 cabins x 4 persons/cabin = 60 persons

60 persons + 12 persons = 72 persons total

The additional 12 persons (20%) is for:

1. Persons using sleeping bags unknown to the park.

2. Persons not staying overnight in the cabins but using the cabin facilities.

Design is based on data from Metcalf and Eddy, <u>Wastewater Engineering</u>, Table 2.11, "Design Unit Sewage Flows for Recreational Facilities," p. 34.

0.17 lbs BOD_5/capita/day or

50 gallons/capita/day

72 persons x 50 = 3600 gallons/day

72 persons x 0.17 = 12.24 lbs BOD_5/day

$$\frac{12.24}{8.34 \times 204 \text{ mg/l}} = 7194 \text{ gallons/day}$$

Proposed treatment plant designed for 7500 gallons/day.

(c) Collection system designed for 3600 gallons/day.

An 8 inch diameter pipe at a slope of 0.60% (velocity of 2.7 feet/second) has a capacity of 612,000 gallons/day ($n = 0.013$).

It is recommended that all collector sewers be 8 inch diameter and service laterals and cleanouts 6 inch diameter.

PROBLEM 3.6

A septic tank system is to be installed for a four bedroom house. Standard trenches are to be used for the sewage disposal system. The soil percolation rate has been found to be 46 inches per minute and deep test pit analyses revealed a water seep was observed at 84 inches below surface. Determine the design of the septic tank leaching area to include trench width and spacing. List any assumptions.

Solution

For three bedrooms the minimum septic tank size is 900 gallons. Add 100 gallons for each additional bedroom. Therefore, design for 1000 gallon septic tank. A percolation rate of 46 minutes per inch allows a sewage application rate of 330 square feet/bedroom.

Required leaching area = 4 x 330 = 1320 ft^2.

Use standard trenches since a steep slope (>15%) is not specified. On steep slopes (maximum 25%) a serial distribution system may be required. If slope is less than 8%, seepage beds may be used. A trench width of 36 inches with minimum 6 inches of crushed stone underneath and minimum of 2 inches stone above perforated pipe is recommended. Each lineal foot of trench can be expected to provide 3 square feet of leaching area, or

$$\frac{1320\ ft^2}{3\ ft^2} = 440 \text{ lineal feet}$$

Recommend 5 laterals each 100 feet long, spaced 9 feet on center. Each line should slope uniformly 2 to 4 inches over the 100 feet. A distribution box should also be installed to insure uniform dosing of each field.

Recommended septic tank absorption area requirements from the State of Pennsylvania Department of Environmental Resources, Chapter 73, Standards for Sewage Disposal Facilities is shown below. This data is for single family residences and includes allowances for garbage grinders and automatic washing machines.

Average Percolation Rates	Septic Tanks (Sq. ft./bedroom)
0 - 5 min/inch	unsuitable
6 - 15 min/inch	175
16 - 30 min/inch	250
31 - 45 min/inch	300
46 - 60 min/inch	330
61 or more min/inch	unsuitable

PROBLEM 3.7

In the operation of a waste stabilization pond,

(a) Explain the function of bacteria and algae.

(b) Why is algae a problem if present in the discharge?

(c) How can algae be controlled?

(d) What is meant by a faculative pond?

Solution

(a) A waste stabilization pond operation is dependent on the reaction of bacteria and algae. Organic matter is metabolized by bacteria to produce the principle products of carbon dioxide, water, and a small amount of ammonia nitrogen. Algae convert sunlight into energy through the process of photosynthesis. They utilize the end products of cell synthesis plus other nutrients to synthesize new cells and produce oxygen. The most important role of the algae is production of oxygen in the pond for use by aerobic bacteria. In the absence of sunlight the algae will consume oxygen the same as bacteria. Algae removal is important in producing a high quality effluent from the pond.

(b) The discharge of algae increases suspended solids in the discharge and may present a problem in meeting water quality criteria. The algae exert an oxygen demand when they settle to the bottom of the stream and undergo respiration.

(c) The following methods have been suggested for control of algae:

(1) Use multiple ponds in series.

(2) Draw off effluent from below the surface by use of a good baffling arrangement to avoid algae concentrations on the surface of the pond.

(3) Use sand filter or rock filter for algae removal.

(4) Use alum addition and flocculation.

(5) Use microscreening.

(6) Effluent chlorination to kill algae. Chlorination may increase BOD loading due to dead algae cells releasing stored organic material.

(d) Faculative ponds have two zones of treatment. An aerobic surface layer in which oxygen is used by aerobic bacteria for waste stabilization and an anaerobic bottom zone in which sludge decomposition occurs. No artificially induced aeration is used.

PROBLEM 3.8

The following conditions have been observed in the activated sludge process:

(a) Excessive white foam on the aeration tank.

(b) Rising clumps of sludge in the final clarifier.

(c) Pinpoint floc observed in the final clarifier.

Outline possible causes and corrective actions.

Solution

(a) White foam on aeration tank:

Causes

(1) Mixed liquor suspended solids concentration in aeration tank too low, F/M ratio too high for organic loading.

(2) Presence of toxic wastes.

(3) Insufficient dissolved oxygen.

Corrective actions

(1) Increase return rate of activated sludge to increase mixed liquor suspended solids concentration.

(2) Check that no toxic industrial wastes are being received at plant.

(3) Increase dissolved oxygen by increasing air rate, possibly modifying air distribution system, or modifying return sludge distribution system.

(b) Rising sludge in final clarifier:

Causes

(1) Excessive nitrification with denitrification in final clarifier causing release of nitrogen gas that floats sludge to surface.

(2) Sludge removal rate is too slow, reducing dissolved oxygen level.

Corrective actions

(1) Increase sludge withdrawal rate from final clarifier.

(2) Increase sludge wasting rate to lower sludge age of system. If possible, reduce oxygen supply to reduce nitrification.

(c) Pinpoint floc in final clarifier:

Causes

(1) Low F/M ratio and high sludge age characteristic of extended aeration operation.

(2) Excessive aeration resulting in dispersed floc formation.

Corrective actions

(1) Gradually increase sludge wasting rate.

(2) Check for proper aeration and mixing in aeration tank.

PROBLEM 3.9

An activated sludge wastewater treatment plant with an average daily flow of 3.5 mgd has an influent phosphorus concentration of 10 mg/l. Stream standards require that the total phosphorus concentration in the plant effluent not exceed 1.0 mg/l. Prepare flow diagrams showing two treatment schemes to meet the effluent limitation. State any special general design considerations for each process identified. What chemical dosage rates would you recommend?

Solution

Two possible treatment schemes to meet the required phosphorus limitation are shown in Figure 3.1 (a) and (b). One method involves adding a metal salt, such as alum or ferric chloride, either to the raw sewage ahead of the primary clarifier, to the aeration tank, or at both locations. The chemical addition should be at a point of rapid mixing. In some plants flocculation with a polyelectrolyte following chemical addition may improve phosphorus removal. Effluent filtration may be required in some cases to consistently meet a total phosphorus limit of 1.0 mg/l. Chemical addition in the raw sewage will result in greater quantities of primary sludge and a decrease in secondary sludge. Anaerobic digestion should essentially be unaffected by alum or ferric chloride unless the digester is overloaded due to greater quantities of waste sludge. Digester alkalinity may be more difficult to control.

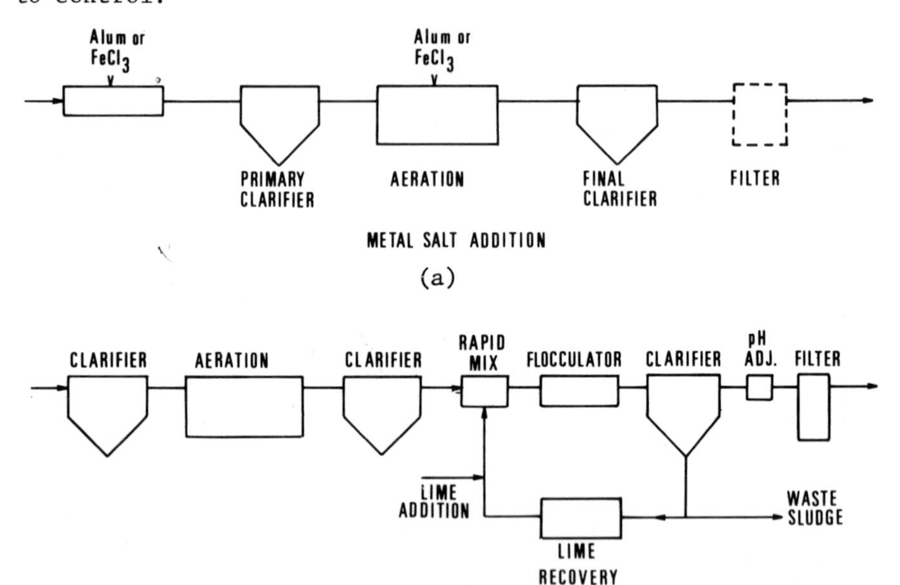

Figure 3.1. Flow schemes for phosphorus removal. (a) Metal salt addition. (b) Single stage lime treatment.

A second method involves lime addition as a tertiary treatment process. A single stage lime treatment system is shown in Figure 3.1 (b). The lime dosage required is primarily dependent on pH and alkalinity. Lime addition to the biological treatment process may not be suitable because of the high pH usually required for phosphorus removal. Lime addition provides good effluent phosphorus control and can be designed to provide an effluent phosphorus level in the range of 0.5 mg/l. Lime sludge may be calcined to recover lime for reuse, but is usually economical only at larger plants.

Because of the variability of wastewater characteristics and plant design, treatment for phosphorus removal should be considered on a case-by-case basis. Jar tests, followed by plant scale tests if possible, are recommended to determine chemical dosage requirements. If alum or ferric chloride is added to the aeration tank mixed liquor, plant tests become more critical because jar tests do not reflect removal capabilities of the activated sludge process. Dewatering characteristics of the sludge may also be evaluated by laboratory tests. Total cost for lime treatment is significantly higher than alum or metal salt addition.

Ratios of alum to phosphorus addition considered reasonably representative for alum treatment of municipal wastewater are shown in Table 3.2.[1] For example, to achieve 95% phosphorus removal from a wastewater containing 10 mg/l, the alum dosage would be,

10 mg/l x 22 = 220 mg/l = 1835 lbs/million gallons

Table 3.2

ALUM ADDITION FOR PHOSPHORUS REMOVAL

Phosphorus Reduction Required	Alum: Phosphorus Weight Ratio
75%	13:1
85%	16:1
95%	22:1

Data in Table 3.3[2] shows alum and ferric chloride addition rates necessary to achieve an effluent total phosphorus concentration of 1.0 mg/l. This data represents a summary of full scale treatability studies and is an average value based on the number of plants surveyed.

Table 3.3

ALUM AND FERRIC CHLORIDE ADDITION FOR PHOSPHORUS REMOVAL

	Addition to Raw Wastewater (Molar Ratio)*	Addition to Mixed Liquor (Molar Ratio)*
Ferric Chloride	2.7:1	1.5:1
Alum	1.7:1	1.6:1

*Average molar ratio of metal ion to total phosphorus

PROBLEM 3.10

List at least three advantages and disadvantages each for a low rate and high rate trickling filter.

Solution

Low Rate

Advantages	Disadvantages
Relatively simple and easy to operate.	Odor problem.
Consistent effluent quality.	Filter flies may breed in filter.
Low power cost since no re-circulation.	Low organic loading compared to high rate filter.
Nitrified effluent.	

High Rate

Advantages	Disadvantages
Continuous food supply for organisms.	Nitrified effluent only at low loading.
Increased BOD removal due to increased contact time.	Added pumping cost.
Increased hydraulic and organic loading possible.	Greater sludge production.
Less odor and fly larvae washed away.	

PROBLEM 3.11

A high rate trickling filter is to be used for a town with population of 5000 and an average daily sewage flow of 90 gallons per capita. Find the required dimensions of the filter necessary to achieve 75% BOD_5 removal without recirculation and at a recirculation ratio of 4:1. Assume influent waste strength to the filter is 150 mg/l BOD_5.

Solution

Solve using National Research Council (NRC) equations.[3] Equations are applicable to both single and multistage filters, with or without recirculation.

$$E = \frac{1}{1 + 0.0085\ (W/VF)^{1/2}}$$

where E = treatment efficiency

W = BOD loading to filter, lbs/day

V = volume of filter media, acre-feet

F = recirculation factor

$$F = \frac{1 + R}{(1 + 0.1R)^2}$$

Calculate filter loading.

$$5000 \times 90 \text{ gpcd} \times 150 \text{ mg/l} \times \frac{1}{10^6} \times 8.34 = 563 \text{ lbs/day}$$

Calculate volume of filter media, V, without recirculation using NRC equation.

$$0.75 = \frac{1}{1 + 0.0085\ (563/V)^{1/2}}$$

$$V = 0.36 \text{ acre-feet}$$

Use a filter depth of 10 feet and calculate filter area and diameter.

$$\text{Area} = \frac{\text{Volume}}{\text{Depth}} = \frac{0.36 \text{ acre-feet}}{10 \text{ feet}} = 1567 \text{ ft}^2$$

Diameter = 45 feet

Calculate volume of filter media with recirculation rate of 4:1.

$$F = \frac{1 + 4}{(1 + 0.1 \times 4)^2} = 2.55$$

$$0.75 = \frac{1}{1 + 0.0085(563/2.55V)^{\frac{1}{2}}}$$

$$V = 0.142 \text{ acre-feet}$$

$$\text{Area} = \frac{0.142 \text{ acre-feet}}{10 \text{ feet}} = 619 \text{ ft}^2$$

Diameter = 29 feet

PROBLEM 3.12

An existing wastewater treatment plant with a flow of 6 mgd is presently removing 70% of the influent 5-day BOD using a low rate trickling filter. The plant must be upgraded to provide a minimum of 92% BOD removal. Outline four alternative methods to meet the required level of treatment.

Solution

Four recommended alternatives would be as follows:

(a) Convert one or more of the trickling filters to aeration basins in order to use the completely mixed activated sludge process. Utilize any remaining trickling filters as roughing filters.

(b) Add new single stage high rate trickling filter with intermediate clarifier ahead of existing low rate filter. Plant is operated as two-stage trickling filter. Add polishing lagoon or filtration as additional treatment step.

(c) Add plastic media roughing filters ahead of existing filter beds. Install recirculation capacity for operation of existing filters as high rate trickling filters. Add additional treatment unit such as polishing lagoon or filtration.

(d) Add rotating biological contactor either as an additional treatment unit following the existing trickling filter or operated in parallel with the existing unit.

For each of the four alternatives described, consideration should be given for additional primary settling tanks, final clarifier, grit removal facilities, and sludge handling facilities necessary for plant upgrading.

PROBLEM 3.13

Two pilot plant trickling filter units are operated in parallel treating the same wastewater under the same operating conditions in order to evaluate two different synthetic plastic media. BOD samples are taken at depths of 5, 10, 15, and 20 feet below the top of the filter. Average results for the test period for three flow rates are shown in Table 3.4. The specific surface area of the media in filter A is 30 sq. ft./cu. ft., the specific surface area for filter B is 27 sq. ft./cu. ft. Based on this test data, determine which filter media provides the better treatment efficiency.

Solution

Problem is solved using equations and method as described by Eckenfelder for trickling filter design.[4] The general design equation for soluble organics removal can be written as,

$$\frac{S_e}{S_o} = e^{-kA_v D^m / Q^n}$$

where S_e = effluent BOD concentration, mg/l

S_o = influent BOD concentration, mg/l

k = removal rate constant

A_v = specific surface, ft^2/ft^3

D = filter depth, feet

Q = hydraulic loading, gpm/ft^2

m,n = constants (assume m = 1.0)

Percent BOD removal versus filter depth is shown in Figure 3.2. The value of n is obtained by plotting the slope obtained from Figure 3.2 versus flow rate. The value of k may then also be graphically determined as shown in Figure 3.3. Data is summarized in Table 3.5.

The media which resulted in the highest k value would be expected to provide the better treatment efficiency since the BOD removal rate is greater. Filter A has a k value of 0.00161, Filter B a k value of 0.00167. Since both of these values are approximately the same, it may be concluded that both filter media will provide approximately the same treatment efficiency based on the given test conditions.

Table 3.4

TEST RESULTS FOR PROBLEM 3.13.

Hydraulic Loading, Q	Depth From Top Of Media (ft.)	Filter A BOD (mg/1)	Filter B BOD (mg/1)
4 gpm/ft^2	Influent	250	250
	5	193	200
	10	138	168
	15	98	138
	20	75	100
2 gpm/ft^2	Influent	250	250
	5	170	163
	10	120	120
	15	83	75
	20	50	50
1 gpm/ft^2	Influent	250	250
	5	150	145
	10	83	88
	15	48	43
	20	28	38

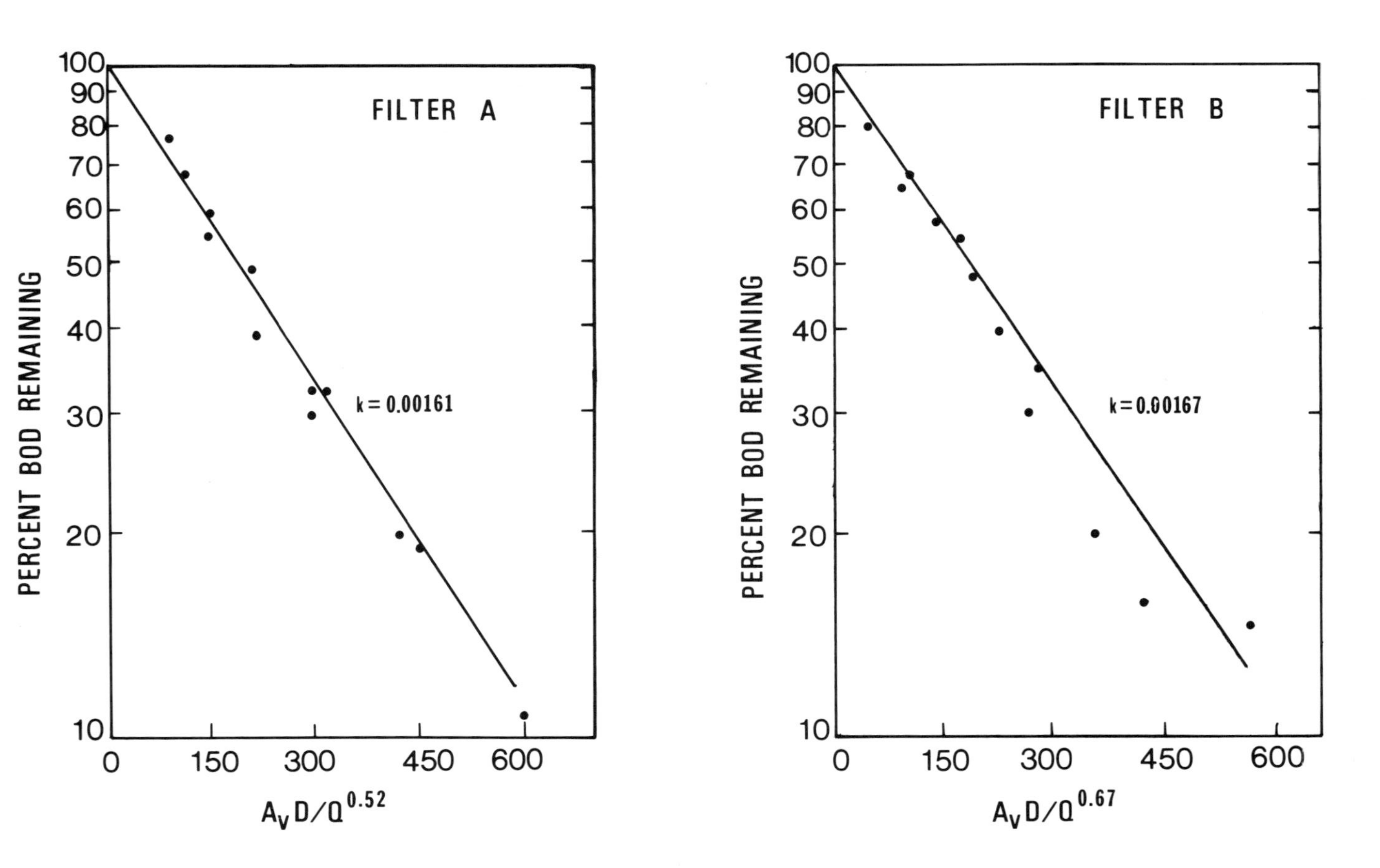

Figure 3.2. Relation between BOD removal and filter depth.

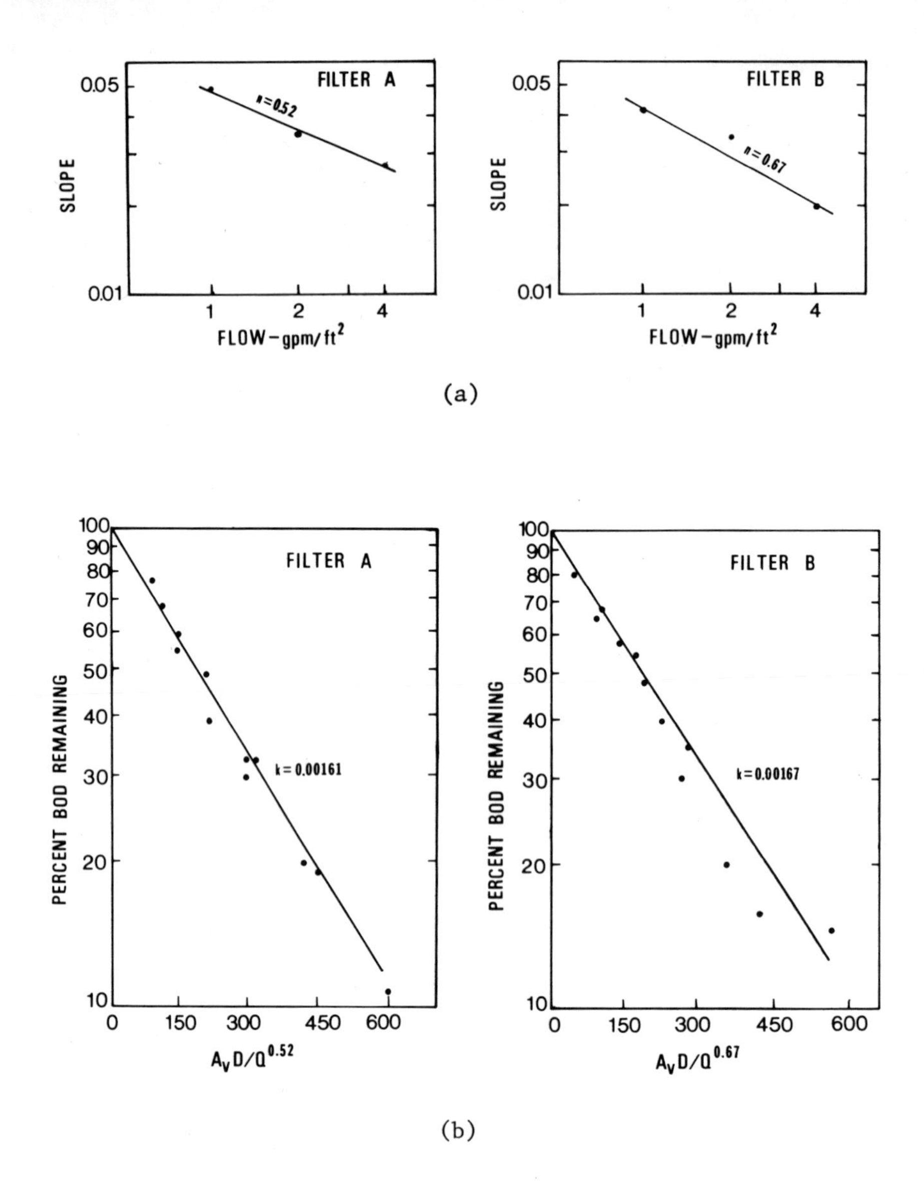

Figure 3.3. Graphical solution for Problem 3.13. (a) Calculation of n value. (b) Calculation of k value.

Table 3.5

SUMMARY OF DATA FOR SOLUTION OF PROBLEM 3.13.

Q	Depth (feet)	Filter A % BOD Remaining	Filter A $A_v\ D/Q^{0.52}$	Filter B % BOD Remaining	Filter B $A_v\ D/Q^{0.67}$
4 gpm/ft^2	5	77	73	80	53
	10	55	146	67	107
	15	39	219	55	160
	20	30	292	40	213
2 gpm/ft^2	5	68	105	65	85
	10	48	209	48	170
	15	33	314	30	255
	20	20	418	20	340
1 gpm/ft^2	5	60	150	58	135
	10	33	300	35	270
	15	19	450	17	405
	20	11	600	15	540

PROBLEM 3.14

A contact stabilization activated sludge process will be designed based on an F/M ratio of 0.3 and a contact tank mixed liquor suspended solids concentration of 3000 mg/l and detention time of 45 minutes. The 5-day BOD loading to the contact tank is 950 pounds/day and suspended solids loading is 700 pounds/day. Design flow rate is 0.8 mgd and a 30% sludge recycle rate may be assumed. Calculate the volume of the contact tank and stabilization tank.

Solution

Calculate contact tank volume for 45 minute detention time.

Volume = detention time x flow

$$V = \frac{45 \text{ min}}{60 \text{ min/hr}} \times \frac{0.8 \times 10^6 \text{ gal}}{24 \text{ hr}} = 25{,}000 \text{ gallons}$$

The F/M ratio may be assumed to be the pounds BOD to contact tank divided by pounds MLVSS in both contact and aeration tanks. Assume MLVSS/MLSS ratio = 0.8.

F/M = 0.3

F = 950 lbs BOD

$$M = \frac{950}{0.3} = 3170 \text{ lbs MLVSS}$$

Calculate MLVSS in contact tank.

3000 mg/l x 0.8 x 0.025 mg x 8.34 = 500 lbs

Stabilization tank solids = 3170 - 500 = 2670 lbs

At 30% sludge recycle rate the total flow to the contact tank = 1.04 mgd.

Make a solids material balance around contact tank to calculate suspended solids concentration in sludge recycle flow from stabilization tank.

Solids in return sludge + influent solids = solids out of contact tank.

Solids in return sludge = (1.04 mgd x 3000 x 8.34) - 700

= 25,320 lbs/day

Calculate return sludge solids concentration for flow of 0.24 mgd.

$$\frac{25{,}320 \text{ lbs/day}}{0.24 \text{ mgd} \times 8.34} = 12{,}650 \text{ mg/l}$$

Calculate stabilization tank volume.

$$\frac{2{,}670 \text{ lbs MLVSS}}{12{,}650 \text{ mg/l} \times 8.34 \times 0.8} = 31{,}600 \text{ gallons}$$

PROBLEM 3.15

Assuming that the following conditions are applicable for the discharge from a sewage treatment plant and for the receiving stream, determine the required percent BOD removal for the plant.

Effluent flow rate: 13.4 mgd

Stream flow rate: 22.1 cfs

Temperature: 25° C

D.O. saturation at 25° C: 8.4 ppm

D.O. of natural stream: 8.0 ppm

D.O. of plant effluent: 2.0 ppm

Minimum required D.O. in stream: 5.0 ppm

K_1 (deoxygenation coefficient): 0.10 @ 20° C, 0.1258 @ 25° C

K_2 (reaeration coefficient): 0.45 @ 20° C, 0.566 @ 25° C

Per capita BOD: 0.23 lb (5 day 20° C)

0.253 lb (5 day 25° C)

0.370 lb (ultimate 25° C)

Solution

Calculate dissolved oxygen below outfall.

$$\frac{(14.3 \text{ mgd} \times 8.0 \text{ mg/l}) + (13.4 \text{ mgd} \times 2.0 \text{ mg/l})}{(14.3 + 13.4)} = 5.10 \text{ mg/l}$$

D_a (initial deficit) = 8.4 - 5.1 = 3.3 mg/l

D_c (critical deficit) = 8.4 - 5.0 = 3.4 mg/l

Use equations from Fair and Geyer, Elements of Water Supply and Waste-Water Disposal.[5]

$f = K_2/K_1 = 0.566/0.1258 = 4.5$

$D_a/D_c = 3.3/3.4 = 0.97$

$L_a/D_c = 4.5$

L_a (allowable load) = 3.4 x 4.5 = 15.3 mg/l ultimate BOD

15.3 x 27.7 mgd x 8.34 = 3540 lb ultimate BOD

Calculate required treatment:

$$\frac{3540 \text{ lbs}}{0.37 \text{ lb/cap} \times 13.4 \text{ mgd}/100 \text{ gpcd}} = 0.071$$

Treatment level = 100% - 7.1% = 92.9%

PROBLEM 3.16

A pilot plant study to determine the settling characteristics of an activated sludge produced the results shown in Table 3.6. Show how this data could be used to size a final clarifier for a maximum sludge recycle rate of 50% of influent flow rate and a mixed liquor suspended solids concentration of 3500 mg/l. Assume plant influent flow rate is 1.0 mgd.

Solution

Settling velocity is plotted versus solids concentration and a straight line is obtained as shown in Figure 3.4. Using this data the mass settling rate or solids flux in lbs/sq. ft. - hr. may be calculated. For example, at a settling rate of 7 ft/hr,

Table 3.6

SLUDGE SETTLING CHARACTERISTICS - PROBLEM 3.16.

Settling Velocity (ft/hr)	Solids Concentration (mg/l)
7.0	2000
4.0	3000
2.0	3700
1.0	5800
0.5	7900

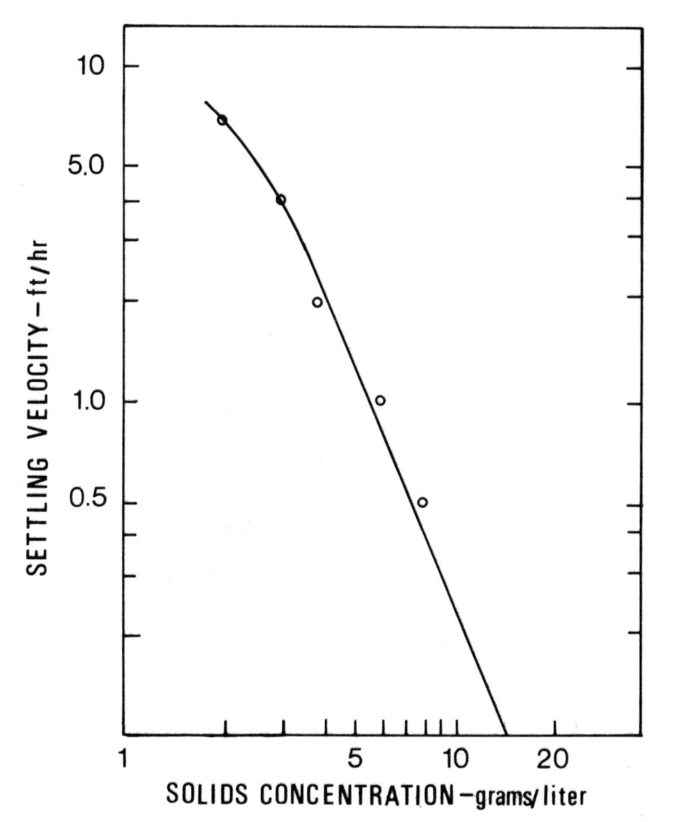

Figure 3.4. Settling velocity versus suspended solids concentration.

$$\text{Solids flux} = \frac{7\ \text{ft}}{\text{hr}} \times \frac{2\ \text{gm/l}}{454\ \text{gm/lb}} \times \frac{28.32\ \text{l}}{\text{ft}^3} \times \frac{24\ \text{hr}}{\text{day}}$$

$$= 21\ \text{lbs/ft}^2 - \text{day}$$

Solids flux can be plotted versus suspended solids concentration as shown in Figure 3.5.

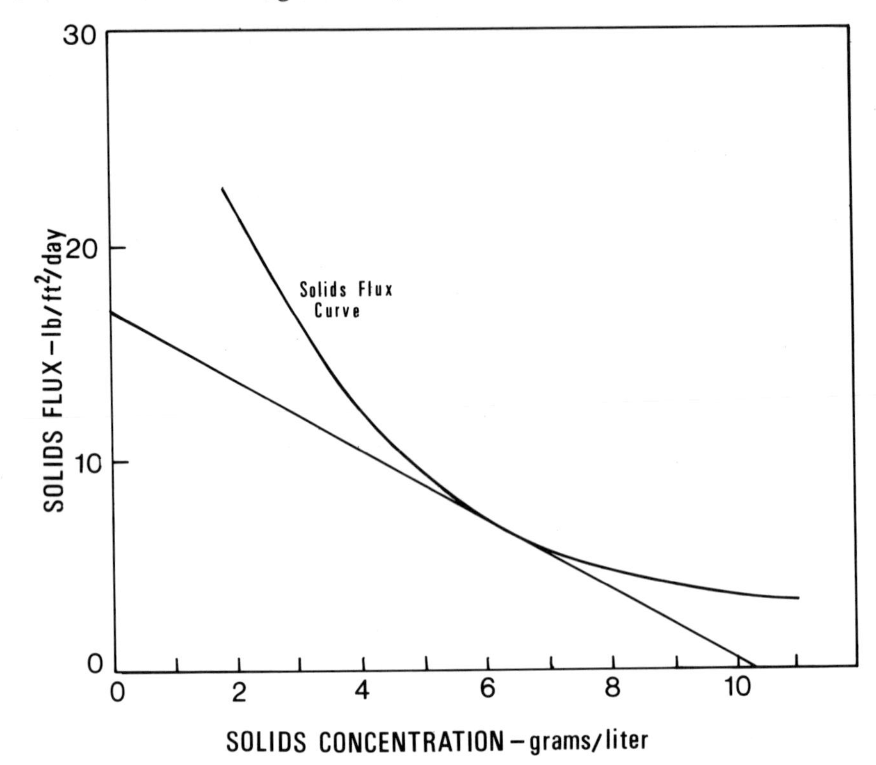

Figure 3.5. Calculation of solids loading from solids flux curve.

Calculate solids loading on clarifier.

[1.0 mgd + 0.5 (1.0 mgd)] (3500 mg/l) (8.34) = 43,785 lbs/day

Calculate return sludge concentration from solids balance around aeration tank. Assume suspended solids concentration from influent waste stream is negligible compared to suspended solids concentration in recycle waste stream.

$$\text{Return sludge concentration} = \frac{1.5\ \text{mgd} \times 3500\ \text{mg/l}}{0.5\ \text{mgd}} = 10{,}500\ \text{mg/l}$$

Using Figure 3.5, draw a line tangent to solids flux curve passing through 10,500 mg/l on x-axis. From y-axis read allowable solids loading of 17 lbs/ft^2/day.

Calculate area required for clarification based on solids loading.

$$\frac{43{,}785 \text{ lb/day}}{17 \text{ lb/ft}^2\text{/day}} = 2576 \text{ ft}^2$$

Calculate area required for clarification based on an assumed overflow rate of 800 gpd/ft^2.

$$\frac{1{,}000{,}000 \text{ gpd}}{800 \text{ gpd/ft}^2} = 1250 \text{ ft}^2$$

Therefore, area required for thickening determines clarifier area.

Use 58 ft. diameter clarifier.

PROBLEM 3.17

An existing primary treatment plant will be upgraded using a pure oxygen activated sludge process. Based on pilot plant studies the design criteria shown in Table 3.7 will be used. Calculate the following:

(a) Aeration volume required.

(b) Oxygen required, pounds/day.

(c) Clarifier area.

Solution

(a) Calculate aeration volume at design flow rate for F/M = 0.55 and MLVSS = 4000 mg/l.

BOD loading = 138 mg/l x 8.34 x 16 mgd = 18,415 lbs/day

$$M = F/0.55 = \frac{18{,}415}{0.55} = 33{,}481 \text{ lbs MLVSS}$$

Table 3.7

DESIGN CRITERIA FOR PROBLEM 3.17.

Primary Effluent Characteristics

Design Flow:	138 mg/l BOD_5
	105 mg/l TSS
Peak Flow:	115 mg/l BOD_5
	95 mg/l TSS

Final Effluent Characteristics

20 mg/l BOD_5

30 mg/l TSS

Average Daily Flow	12 mgd
Design Flow	16 mgd
Peak Daily Flow	20 mgd

F/M ratio 0.55 lbs BOD applied/lb MLVSS

MLVSS 4000 mg/l

MLVSS/MLSS ratio 0.78

Recycle ratio at design flow 33%

Oxygen uptake rate 30 mg/l - hr/gm MLVSS

Clarifier Solids Loading

35 lbs/ft^2 - day at design flow

45 lbs/ft^2 - day at peak daily flow

Clarifier Hydraulic Loading

800 gpd/ft^2 at design flow

1000 gpd/ft^2 at peak daily flow

$$\text{Volume} = \frac{33{,}481}{8.34 \times 4000} = 1.0 \text{ million gallons}$$

Check detention time at design flow and peak daily flow.

$$\frac{1.0 \times 10^6 \text{ gal}}{16 \times 10^6 \text{ gal/24 hr}} = 1.5 \text{ hours @ design flow}$$

$$\frac{1.0 \times 10^6 \text{ gal}}{20.0 \times 10^6 \text{ gal/24 hr}} = 1.2 \text{ hours @ peak daily flow}$$

Detention times appear reasonable and can be compared with pilot plant data.

(b) Calculate area required for final clarifier

For thickening,

Flow = Q + R = 16 mgd + (0.33 x 16) = 21.3 mgd

$$\text{MLSS} = \frac{4000 \text{ mg/l}}{0.78} = 5128 \text{ mg/l}$$

Solids loading = 5128 mg/l x 8.34 x 21.3

= 910,950

$$\frac{910{,}950 \text{ lbs/day}}{35 \text{ lbs/ft}^2\text{-day}} = 26{,}000 \text{ ft}^2 \text{ clarifier area}$$

For clarification,

$$\frac{16 \times 10^6 \text{ gpd}}{800 \text{ gpd/ft}^2} = 20{,}000 \text{ ft}^2 \text{ clarifier area}$$

Choose 26,000 ft^2 based on thickening requirement.

From solids balance around aeration tank return soli concentration (RSC) is calculated as 21,000 mg/l for recycle ratio (R) of 33%.

$$\text{RSC} = \frac{-(16 \times 105 \text{ mg/l}) + (21.3 \times 5128 \text{ mg/l})}{0.33 \times 16}$$

RSC = 21,000 mg/l

Design is now evaluated at peak daily flow conditions Assume return solids concentration and recycle flow remains constant. Calculate F/M ratio and MLSS conc tration at peak flow.

$$\text{MLSS} = \frac{R/Q}{1 + R/Q} \times \text{RSC}$$

$$R/Q = \frac{0.33 \times 16}{20} = 0.264$$

$$\text{MLSS} = \frac{0.264}{1.264} \times 21{,}000 = 4386 \text{ mg/l}$$

$$\text{F/M} = \frac{20 \text{ mgd} \times 115 \text{ mg/l}}{4386 \text{ mg/l} \times 0.78 \times 1.0 \text{ mg}} = 0.67$$

F/M ratio is considered within acceptable operating range.

Check area required for final clarifier at peak daily flow.

For thickening,

$Q + R = 20 \text{ mgd} + (0.264 \times 20) = 25.3 \text{ mgd}$

Solids loading = 4386 mg/l x 8.34 x 25.3 mgd
= 925,450 lbs/day

$$\frac{925{,}450 \text{ lbs/day}}{45 \text{ lbs/ft}^2\text{-day}} = 20{,}500 \text{ ft}^2 \text{ clarifier area}$$

For clarification,

$$\frac{20 \times 10^6 \text{ gpd}}{1000 \text{ gpd/ft}^2} = 20{,}000 \text{ ft}^2 \text{ clarifier area}$$

(c) Calculate oxygen requirements based on oxygen uptake rate of 30 mg/l-hr/gm MLVSS.

30 x 0.001 x 4000 x 1.0 x 8.34 x 24 = 24,019 lbs/day

Assuming 90% pure oxygen dissolves.

$$\frac{24{,}019}{0.9} = 26{,}690 \text{ lbs Oxygen/day required}$$

Design for 50% above required level or 40,000 lbs/day oxygen.

Design Summary

Aeration volume: 1.0 million gallons

Clarifier area: 26,000 ft^2

Oxygen supply: 40,000 lbs/day

PROBLEM 3.18

The data shown in Table 3.8 is collected in a non-steady state oxygen transfer test using tap water. Water temperature is 17° C and oxygen saturation value is 9.51 mg/l.

Table 3.8

TEST DATA FOR PROBLEM 3.18.

Time (Minutes)	Dissolved Oxygen (mg/l)
0	1.00
0.25	2.00
0.50	2.95
0.75	3.88
1.00	4.65
1.50	5.75
2.00	6.65
2.50	7.30
3.00	7.82
3.50	8.20
4.00	8.55

The test is repeated under steady state conditions using an equivalent volume of the wastewater to be tested. Other test conditions remain the same. Test results are as follows:

Oxygen saturation concentration = 9.15 mg/l

Oxygen equilibrium concentration = 5.85 mg/l

Oxygen uptake rate = 51.2 mg/l/hr

(a) Calculate the oxygen transfer coefficient, $K_L a$, for the tap water and the wastewater.

(b) Calculate the value of Alpha (α) and Beta (β).

Solution

(a) For non-steady state conditions the rate of oxygen transfer is given by the equation,

$$\frac{dC}{dt} = K_L a \ (C_s - C)$$

The equation may be integrated and rewritten as,

$$K_La = \frac{2.303 \log_{10} [(C_s - C_1)/(C_s - C_2)]}{t_2 - t_1}$$

where K_La = transfer coefficient

C_s = saturation concentration, mg/l

C_1, C_2 = dissolved oxygen concentration at times t_1 and t_2

The data from Table 3.8 is plotted in Figure 3.6 and K_La calculated as

$$K_La = \frac{2.303 \log_{10} [(9.51 - 1.0)/(9.51 - 8.20)]}{3.5 \text{ minutes}}$$

$$K_La = 0.53 \text{ minutes}^{-1}$$

For steady state conditions K_La may be calculated from the equation,

$$\frac{dC}{dt} = r = K_La \ (C_s - C)$$

At steady state dC/dt = 0 and the equation may be written as,

$$K_La = \frac{r}{C_s - C_L}$$

where r = oxygen uptake rate, mg/l/hr

C_s = oxygen saturation concentration, mg/l

C_L = oxygen equilibrium concentration, mg/l

$$K_La = \frac{51.2}{9.15 - 5.85} = 15.5 \text{ hour}^{-1}$$

$$K_La = 0.258 \text{ minutes}^{-1}$$

(b) $\text{Alpha} = \dfrac{K_La \text{ waste water}}{K_La \text{ tap water}}$

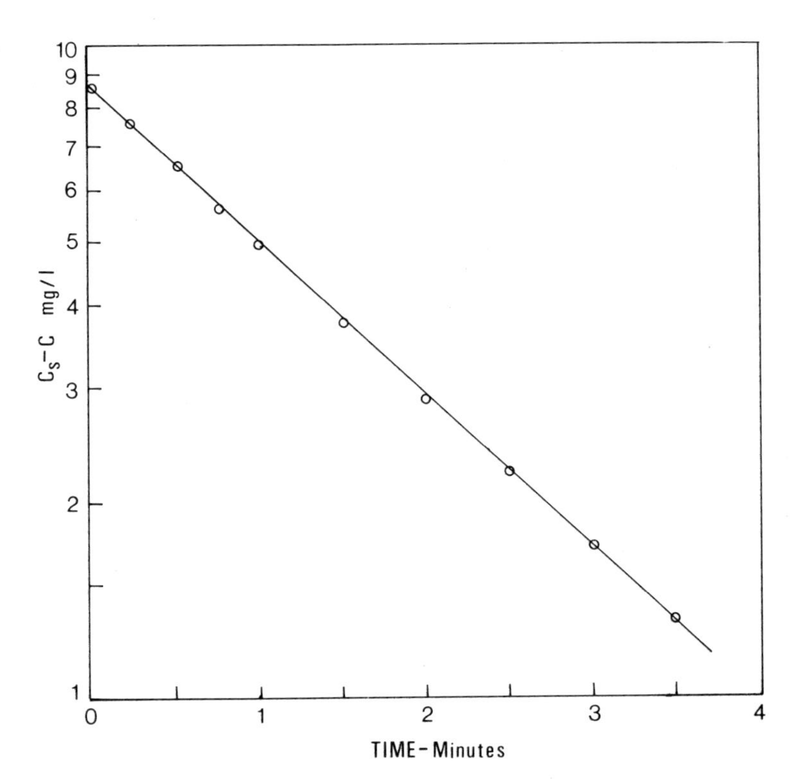

Figure 3.6. Oxygen Uptake Rate for Problem 3.18.

Alpha = 0.258/0.530 = 0.487

$$\text{Beta} = \frac{\text{oxygen saturation waste water}}{\text{oxygen saturation tap water}}$$

Beta = 9.15/9.51 = 0.962

PROBLEM 3.19

An aeration equipment manufacturer has recommended installation of a mechanical surface aerator with a clean water transfer rate of 3.0 lbs O_2/hp — hr. The following additional information is also available:

Dissolved oxygen required in effluent = 2.0 mg/l

Minimum wastewater temperature = 24° C

Maximum wastewater temperature = 35° C

Plant elevation = 3000 feet above sea level

$\alpha = 0.9 \qquad \beta = 0.95$

The oxygen uptake rate has been determined as 28 mg/l-hr. Aeration basin volume is 4.1 million gallons. Calculate the annual cost of operation if electricity is 5¢ per kw-hr.

Solution

Calculate field transfer rate, N, using the equation,

$$N = N_o \frac{\beta \times (C_{sat} - C_L)}{C_{SC}} \quad 1.024^{T-20} \alpha$$

where N_o = clean water transfer rate

C_{SAT} = saturation concentration at design temperature and altitude = 6.4 mg/l

C_L = Dissolved oxygen level of aeration basin

C_{SC} = saturation concentration at sea level and 20° C = 9.2 mg/l

T = wastewater temperature, °C

Assume dissolved oxygen of aeration tank equals dissolved oxygen required in effluent. Maximum wastewater temperature is used for design.

$$N = 3.0 \frac{0.95 \times (6.4 - 2.0)}{9.2} \times 1.024^{35-20} \times 0.9$$

$$N = 1.75 \text{ lb } O_2/\text{hp-hr}$$

Calculate pounds oxygen required.

4.1 million gallons x 28 mg/l/hr x 24 hr/day x 8.34

= 23,000 lb/day

Calculate horsepower required.

$$\frac{23{,}000 \text{ lbs/day}}{24 \times 1.75 \text{ lbs } O_2/\text{hp-hr}} = 548 \text{ hp}$$

Calculate cost for 365 days/year operation and 94% motor efficiency (assumed).

$$\frac{548 \text{ hp}}{0.94} \times 0.746 \frac{\text{kw}}{\text{hp}} \times \frac{\$0.05}{\text{kw-hr}} \times 24 \times 365 = \$190{,}500/\text{year}$$

PROBLEM 3.20

Solids retention time is to be used as an operational control technique in an activated sludge plant with an average daily flow rate of 8 mgd. Aeration tank volume is 2 million gallons. Develop an operating diagram to show the relationship between solids retention time and sludge wasting rate.

Solution

Solids retention time (SRT) can be defined by the following equation,

$$SRT = \frac{\text{pounds of biological solids under aeration}}{\text{pounds of solids removed from the system per day}}$$

SRT can therefore be controlled by the amount of sludge wasted from the system.

For an activated sludge system with solids recycle from the final clarifier, SRT may be written as,

$$SRT = \frac{VX}{X_e\ (Q - Q_w) + Q_w X_r} \qquad \text{(Eqn. 1)}$$

where V = aeration tank volume, gallons

X = mixed liquor suspended solids (or mixed liquor volatile suspended solids) concentration, mg/l

X_e = effluent suspended solids concentration, mg/l

Q = plant flow rate, gallons/day

Q_w = waste sludge flow rate, gallons

X_r = return sludge suspended solids concentration, mg/l

If the solids in the plant effluent, $X_e\ (Q - Q_w)$, is considered negligible compared to the mass of solids in the aeration tank or return sludge stream, equation (1) may be rewritten as:

$$SRT = \frac{VX}{Q_w X_r} \qquad \text{(Eqn. 2)}$$

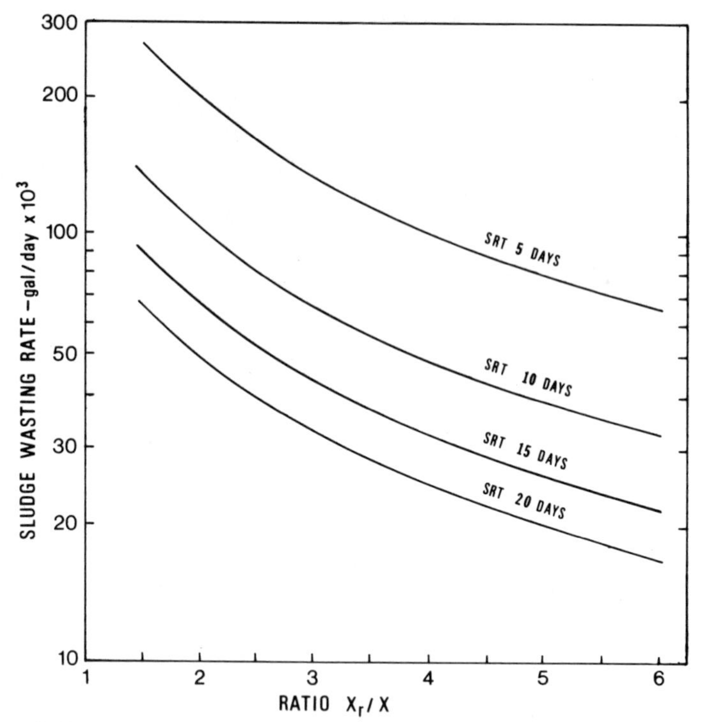

Figure 3.7. Operating diagram for activated sludge plant using solids retention time.

Using equation (2), Figure 3.7 shows the sludge wasting rate plotted versus the ratio X_r/X for SRT values ranging from 5 to 20 days. By knowing the ratio of the aeration mixed liquor suspended solids and the solids concentration in the return sludge, the sludge wasting rate can be adjusted for the SRT value desired for plant operation.

PROBLEM 3.21

Operating data from a waste treatment plant is shown in Table 3.9. Each operating period may be assumed to represent steady-state operating conditions. The plant operates as a complete mix activated sludge process with return sludge recycle from the final clarifier to the aeration tank. Using kinetic equations by Lawrence and McCarty,[6] calculate the following:

(a) The growth yield coefficient and the endogenous respiration rate coefficient.

(b) Plant effluent BOD_5 concentrations as a function of solids retention time.

Solution

For a complete mix system with recycle, the Lawrence and McCarty equations for steady-state operation can be summarized as follows:

$$\frac{1}{SRT} = Y \frac{F}{M} - b \qquad \text{(Eqn. 1)}$$

$$\frac{F}{M} = \frac{k\ S_e}{K_s + S_e}$$

$$\text{or} \quad \frac{1}{F/M} = \frac{K_s}{k} \frac{1}{S_e} + \frac{1}{k} \qquad \text{(Eqn. 2)}$$

$$S_e = \frac{K_s\ [1 + b(SRT)]}{SRT\ (Yk - b) - 1} \qquad \text{(Eqn. 3)}$$

For this particular problem, these terms may be defined as:

SRT = solids retention time, days

Y = yield coefficient, lbs biological solids produced (MLVSS)/lb soluble BOD_5 removed

$\frac{F}{M}$ = food to mass ratio, lbs soluble BOD_5 removed/lb mixed liquor volatile suspended solids (MLVSS).

b = endogenous respiration coefficient, day^{-1}

k = maximum substrate removal rate, lbs soluble BOD_5 removed/day per lb MLVSS

K_s = BOD_5 concentration at a substrate removal rate of k/2, mg/l

S_e = soluble effluent BOD_5 concentration

The F/M ratio is calculated from the equation:

$$\frac{F}{M} = \frac{Q\ (S_o - S_e)}{VX}$$

Table 3.9

PLANT OPERATING DATA, PROBLEM 3.21.

Operating Period	Flow Rate (mgd)	SRT (Days)	Soluble BOD_5 (mg/l) Primary Effluent	Soluble BOD_5 (mg/l) Final Effluent (S_e)	TSS (mg/l) Final Effluent	MLVSS (mg/l)
1	1.24	11.8	160	26	36	1916
2	1.06	10.7	200	24	28	2340
3	0.80	13.0	239	23	23	2120
4	0.90	8.5	253	33	45	2310
5	1.08	15.3	195	25	39	2625
6	1.09	8.2	297	39	54	2664
7	1.26	9.5	226	24	28	2190
8	1.03	9.8	230	30	34	1794
9	1.09	9.0	240	30	32	2343
10	0.75	14.1	247	23	27	2461

Footnote: Aeration volume = 800,000 gallons

where Q = flow rate, mgd

S_o = soluble BOD_5 concentration from primary clarifier, mg/l

S_e = soluble plant effluent BOD_5 concentration, mg/l

V = volume of aeration tank, million gallons

X = MLVSS concentration under aeration, mg/l

Calculated values of 1/SRT, F/M, 1/(F/M), and $1/S_e$ are given in Table 3.10.

Table 3.10

SUMMARY OF CALCULATED DATA FOR PROBLEM 3.21.

Operating Period	$\frac{1}{SRT}$	F/M	$\frac{1}{F/M}$	$\frac{1}{S_e}$
1	0.085	0.108	9.3	0.038
2	0.093	0.100	10.0	0.042
3	0.077	0.100	10.0	0.043
4	0.118	0.116	8.6	0.030
5	0.065	0.083	11.4	0.040
6	0.122	0.132	7.6	0.026
7	0.105	0.146	6.8	0.042
8	0.102	0.144	6.9	0.033
9	0.111	0.122	8.2	0.033
10	0.071	0.086	11.6	0.043

(a) The yield rate and endogenous coefficient are determined by plotting 1/SRT versus F/M as shown in Figure 3.8. From equation 1 the slope of this line is the yield rate, Y, and the y-intercept the endogenous coefficient, b. Y is determined to be 1.2 lbs MLVSS/lb soluble BOD removed; b = 0.035 $days^{-1}$.

Figure 3.9 shows a graph of 1/(F/M) versus $1/S_e$. From equation 2, 1/k = 1.7, k = 0.6 $days^{-1}$. K_s/k = 215, K_s = 129 mg/l.

(b) Figure 3.10 is a graph of effluent BOD versus solids

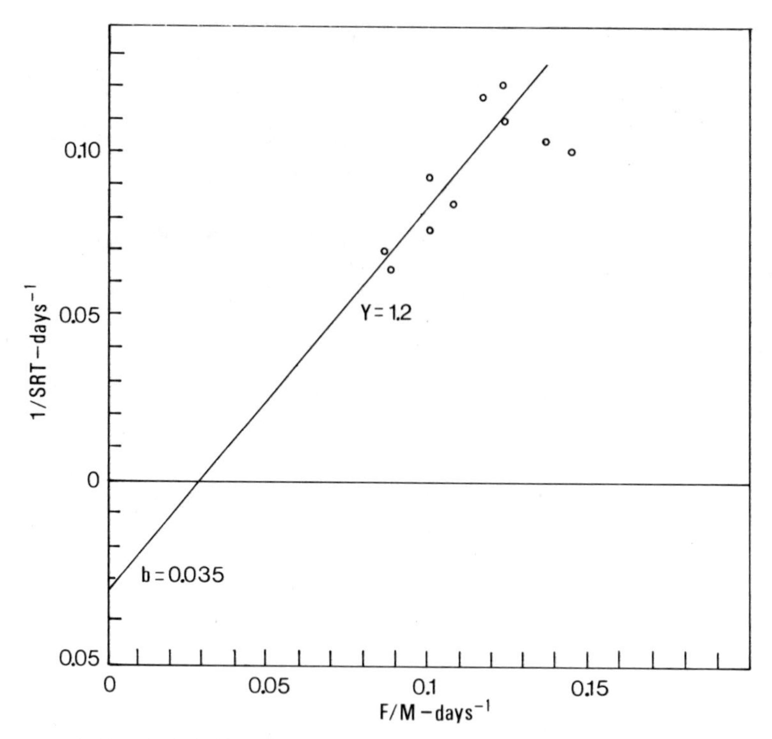

Figure 3.8. Yield and endogenous coefficients.

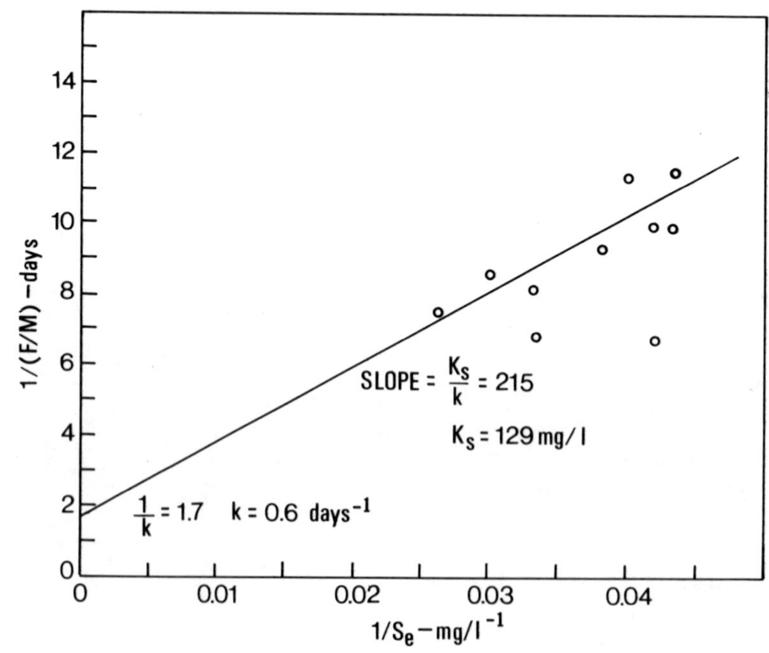

Figure 3.9. Determination of coefficients K_s and K.

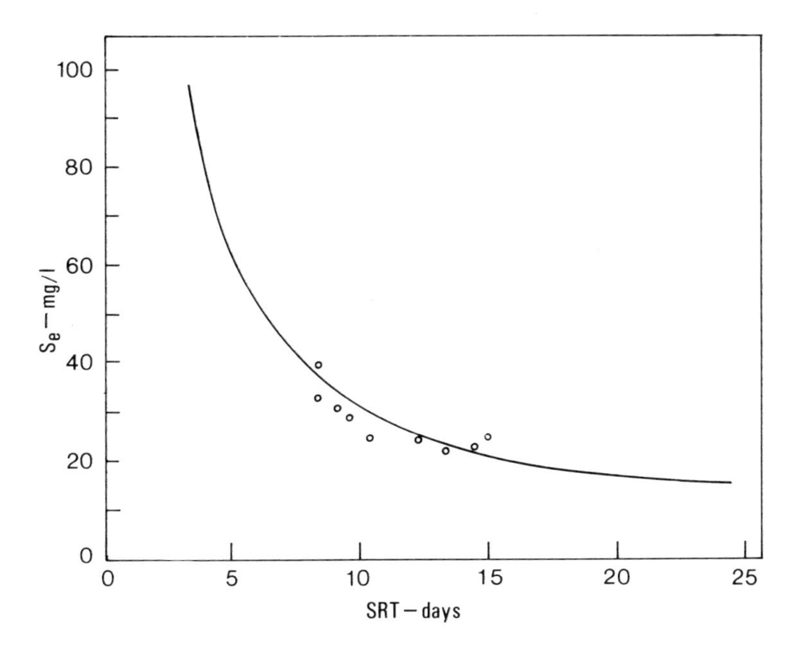

Figure 3.10. Effluent BOD (S_e) as a function of solids retention time.

retention time using equation 3 and values of K_s, k, Y, and b using Figures 3.8 and 3.9. Actual values of S_e from Table 3.9 are also shown on Figure 3.10. Actual data shows good correlation with predicted values.

From Figure 3.10 it becomes apparent that operating at a solids retention time greater than approximately 15 days will result in no significant improvement in effluent quality. At a solids retention time of approximately 3 to 4 days, biological solids are "washed out" of the system and effluent quality deteriorates. The optimum SRT operating range will be dependent on actual operating conditions, and kinetic coefficients may be expected to change with variables such as temperature and wastewater composition. Figure 3.10 indicates it may be desirable to operate at a high solids retention time to improve treatment efficiency and reduce waste sludge quantity and reduce operating costs. However, a point is reached in the system where the increased solids loading on the final clarifier will become a limiting factor.

In general, for an activated sludge treatment process, values of K_s and k may be expected to double with each 10° C increase in temperature. The expected loss of treatment efficiency at lower temperatures may be minimized to some extent by operating at a higher solid retention time. For example, at an SRT of 10 days, equation (3) predicts an effluent BOD of 30 mg/l. If values of K_s and k are reduced 50% to 65 mg/l and 0.3 days^{-1}, respectively, then the solids retention time would need to be increased to 13 days to achieve an effluent BOD of 30 mg/l.

PROBLEM 3.22

Consider the design criteria given in Problem 3.2. Use the Lawrence and McCarty equations[7] to calculate the required aeration tank volume.

Solution

Assume that the following kinetic coefficients are applicable for treatment of domestic wastewater:

$$Y = 0.65$$

$$b = 0.05 \text{ day}^{-1}$$

Assume F/M ratio = 0.3 and calculate value for SRT,

$$\frac{1}{SRT} = Y \frac{F}{M} - b$$

$$= 0.65\ (0.3) - 0.05$$

$$SRT = 7 \text{ days}$$

From Problem 3.2, BOD loading to aeration tanks is 3700 lbs/day, or 139 mg/l at a flow rate of 3.2 mgd. Aeration volume may be calculated for an SRT value of 7 days from the equation,

$$XV = \frac{Y\ (S_o - S_e)\ SRT}{1 + b\ (SRT)} \quad Q$$

where S_o = influent BOD_5 concentration, mg/l

S_e = effluent BOD_5 concentration, mg/l

Q = flow rate, mgd

X = MLVSS concentration, mg/l

V = aeration volume, million gallons

$$V\ (2750)(0.8) = \frac{0.65\ (139 - 30)(7)}{1 + 0.05\ (7)} \times 3.2$$

$$V = 0.53 \text{ million gallons} = 71{,}000 \text{ ft}^3$$

PROBLEM 3.23

A pilot plant study is conducted to determine the feasibility of treating a high strength industrial waste. Table 3.11 summarizes the results of a pilot scale complete mix activated sludge system with solids recycle. Table 3.12 shows results of a system operated as an aerated lagoon. The data is representative of six operating periods at varying F/M ratios. Evaluate the following design criteria for each treatment process: BOD removal rate, treatment efficiency, oxygen utilization rate, yield coefficient, and endogenous respiration coefficient.

Solution

A summary of the data used to develop Figures 3.11 through 3.15 is given in Table 3.13. The following nomenclature is used:

k = BOD_5 removal rate, 1/mg-day

F/M = food/microorganism ratio, $days^{-1}$

Y = cell yield coefficient

b = endogenous respiration coefficient, $days^{-1}$

Y' = oxygen utilization coefficient for cell synthesis, lbs oxygen utilized/lb organics removed

b' = oxygen utilization coefficient for endogenous respiration, lbs oxygen utilized/day per lb MLVSS

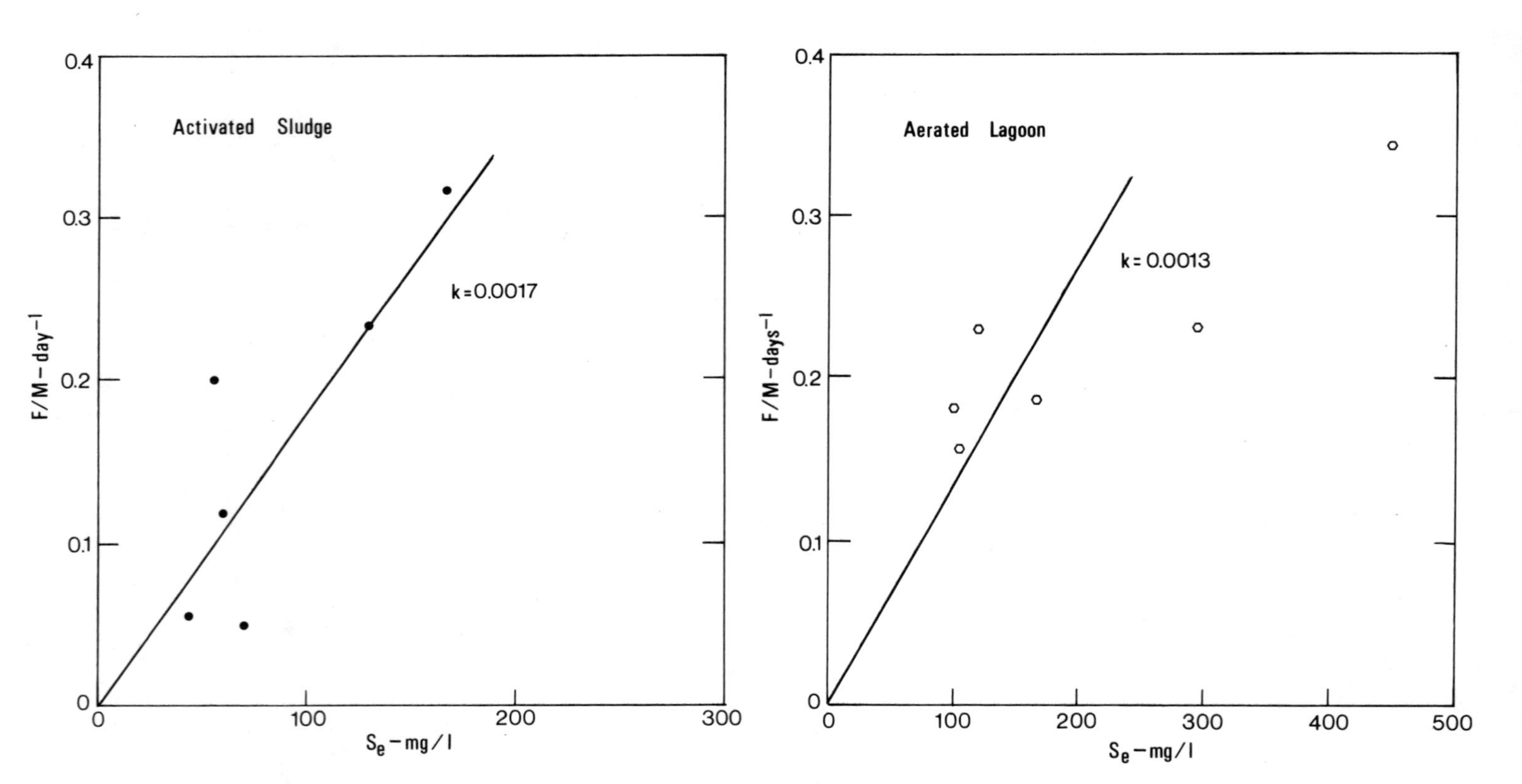

Figure 3.11. BOD removal rate for activated sludge and aerated lagoon.

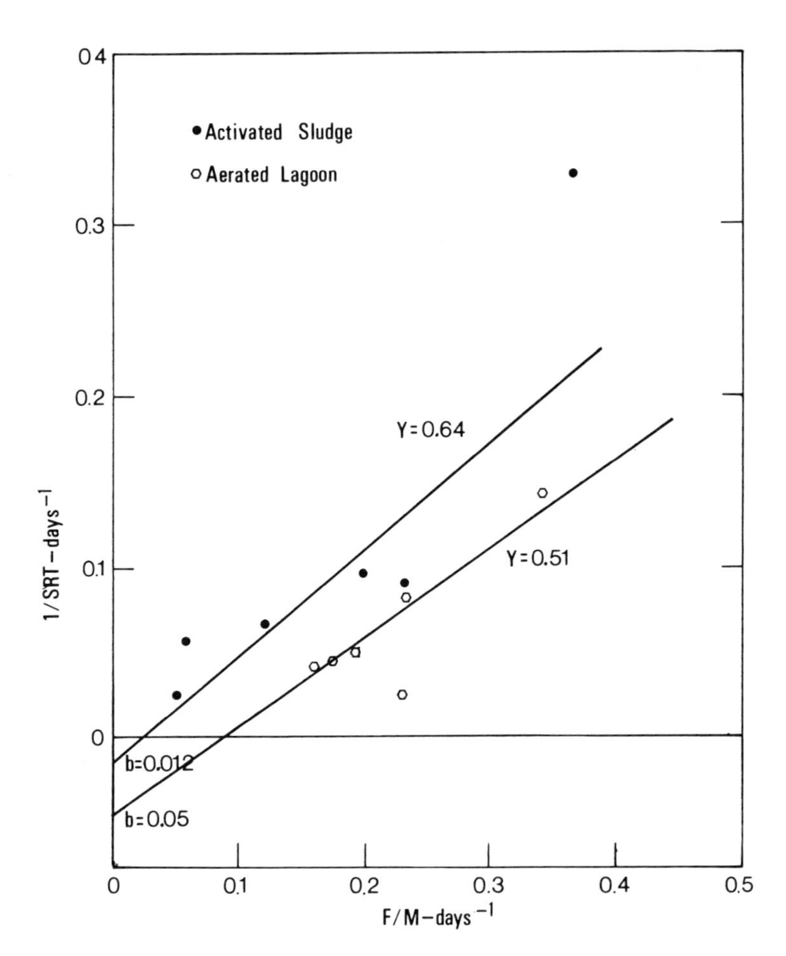

Figure 3.12. Determination of yield coefficient, Y, and endogenous coefficient, b.

SRT = solids retention time, days

S_e = effluent soluble BOD_5, mg/l

Note that in Figures 3.11, 3.12, and 3.13 the F/M ratio is calculated as pounds BOD removed (influent total BOD minus effluent soluble BOD) divided by pounds MLVSS under aeration. In Figure 3.14, the F/M ratio is calculated as pounds BOD applied divided by pounds MLVSS under aeration.

The BOD removal rate coefficient, k, is determined graphically in Figure 3.11 by plotting the F/M ratio versus soluble effluent BOD. Determination of the yield coefficient, Y, and endogenous respiration coefficient, b, is shown in Figure 3.12.

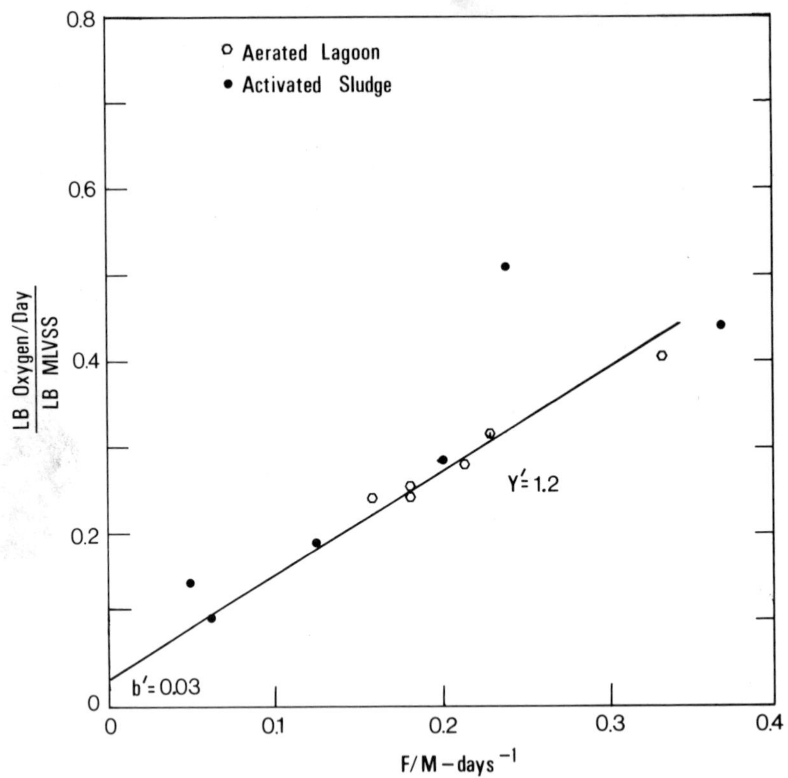

Figure 3.13. Determination of oxygen utilization coefficients.

The oxygen requirements for each system may be defined by the equation,

$$\text{lbs } O_2 \text{ required/day} = Y' \text{ (lbs organics removed/day)} + b' \text{ (lbs MLVSS)}$$

The coefficients Y' and b' are determined from Figure 3.13. Y' is calculated as the slope of the line and b' as the y-intercept. For this example, both systems are determined to have the same coefficients.

When evaluating pilot plant results for an activated sludge system, it is important to determine the optimum F/M ratio and solids retention time for both BOD removal and sludge settleability in the final clarifier. Sludge settling properties at varying F/M ratios can be determined based on the zone settling velocity or sludge volume index. Operation of the final clarifier is critical to good treatment efficiency. At a high F/M ratio (low SRT) dispersed or filamentous sludge can result in a bulking sludge with poor

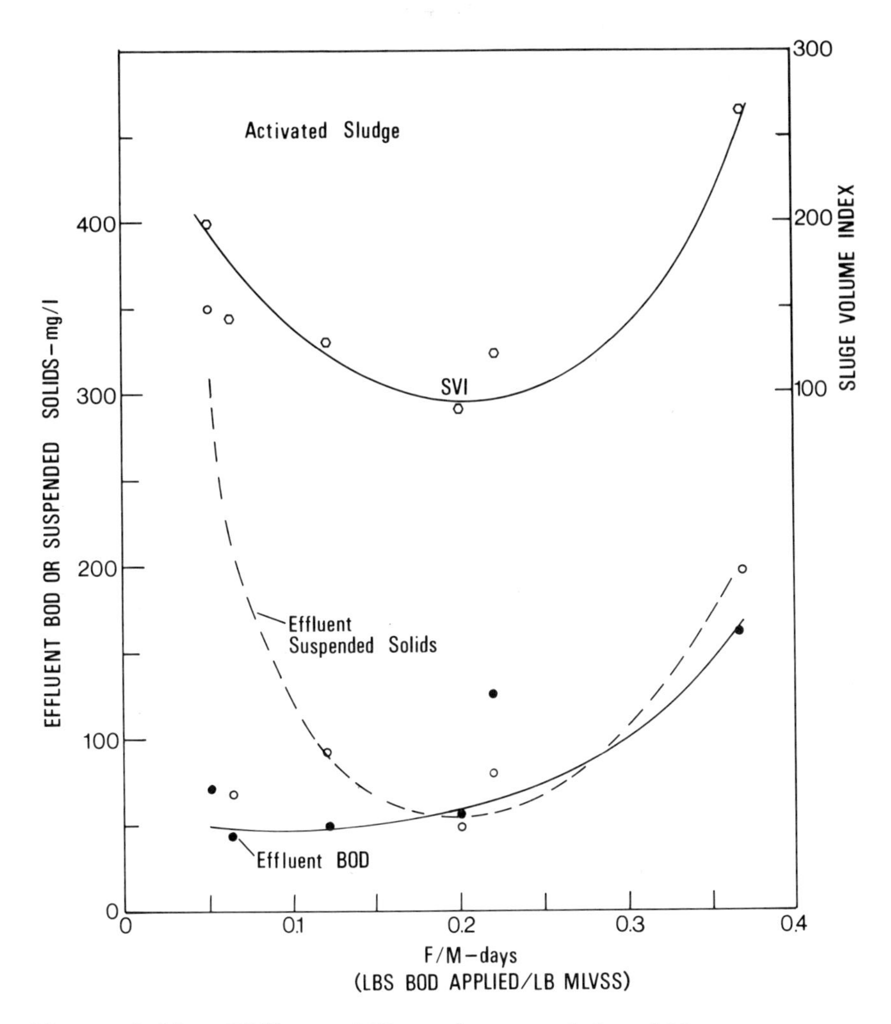

Figure 3.14. Effluent BOD_5 and suspended solids as a function of F/M ratio.

settling characteristics. At too low an F/M ratio (high SRT) a dispersed or pin point floc may result in solids carry over in the final effluent.

For this example, Figure 3.14 shows that the optimum F/M ratio would be in the range of 0.15 to 0.25. Good sludge settling properties with low effluent suspended solids are characteristic of this operating range. Operation at a F/M ratio of 0.25 would probably be more desirable to insure good sludge settleability. Figure 3.15 shows that the optimum SRT would be in the range of 10 to 15 days. At an SRT of less than approximately 5 to 8 days, poor BOD removal

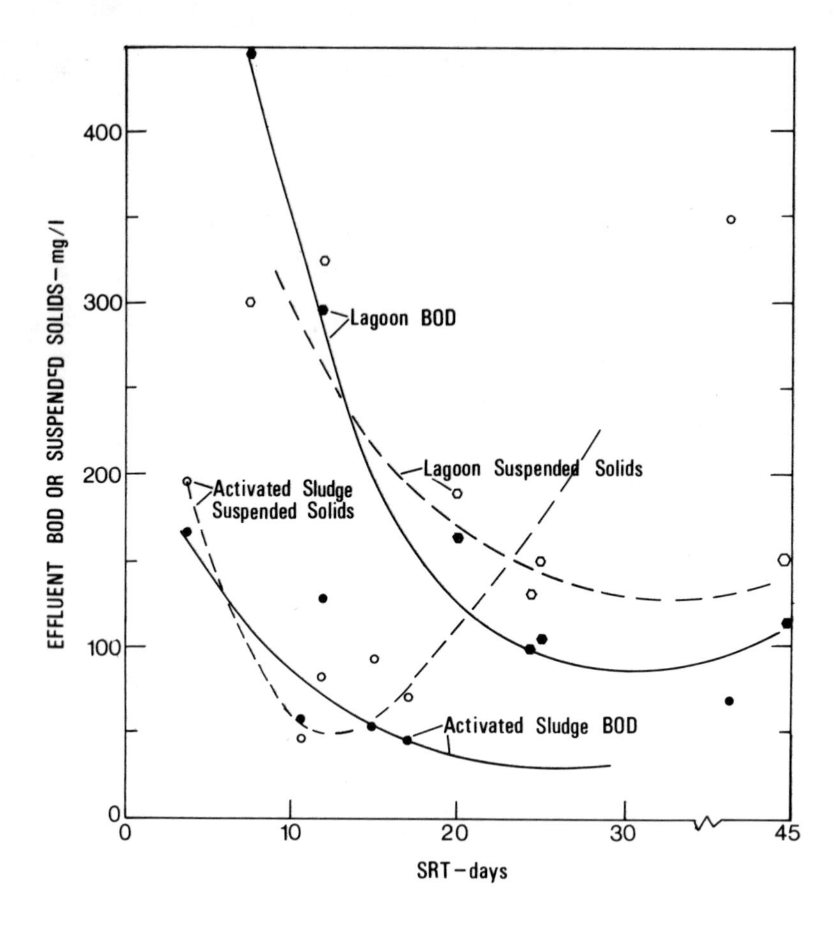

Figure 3.15. Effluent BOD_5 and suspended solids as a function of solids retention time (SRT).

efficiency would be expected.

Aerated lagoons are designed without solids recycle at a lower MLVSS than the activated sludge process. A longer solids retention time is needed to achieve satisfactory BOD removals. Aerated lagoons may be designed as two stage systems using one or more completely mixed aerated lagoon(s) followed by a faculative lagoon with a lower power level to allow solids settling and lower effluent suspended solids. For a completely mixed lagoon, aerator horsepower for adequate mixing to maintain solids in suspension as well as for oxygen transfer must be provided.

Table 3.11

ACTIVATED SLUDGE TEST DATA

	No. 1	No. 2	No. 3	No. 4	No. 5	No. 6
Flow liters/day	6.43	12.8	25.36	45.3	47.0	97.5
Influent BOD_5 mg/1	2013	2013	2013	2013	1989	1989
Effluent total BOD_5 mg/1	98	48	70	70	174	440
Effluent soluble BOD_5 mg/1	69	43	53	55	129	165
Effluent suspended solids mg/1	352	70	92	49	80	190
MLSS, mg/1	2270	3521	3640	3995	3318	4138
MLVSS, mg/1	1944	3060	3163	3491	2936	3776
Oxygen uptake rate mg/1-hr	11.7	13.8	24.1	41.4	63.2	69.2
Temperature °C	18	18	17	18	17	17
Hydraulic Detention Time Days	19.7	9.9	5.0	2.8	2.7	1.3
Solids Retention Time Days	40	17	15	11	11.5	3.0
Sludge Volume Index	200	140	130	90	125	270

Table 3.12

AERATED LAGOON TEST DATA

	No. 1	No. 2	No. 3	No. 4	No. 5	No. 6
Flow liters/day	17.9	10.0	5.1	5.17	2.8	6.34
Influent BOD_5 mg/1	2013	2013	2013	2087	2013	2085
Effluent total BOD_5 mg/1	690	420	156	155	161	200
Effluent soluble BOD_5 mg/1	449	294	106	100	115	165
Effluent suspended solids mg/1	300	320	150	135	155	190
MLSS, mg/1	760	685	555	502	210	610
MLVSS, mg/1	650	589	472	451	184	520
Oxygen uptake rate mg/1-hr	21.7	13.5	9.2	9.2	4.7	10.4
Temperature °C	18	18	18	18	17	17
Hydraulic detention time Days	7.1	12.6	25.0	24.5	45.1	20
Solids Retention Time Days	7.1	12.6	25.0	24.5	45.1	20

Table 3.13

SUMMARY OF DATA ANALYSIS. PROBLEM 3.23

	Activated Sludge Units						Aerated Lagoon Units					
	No. 1	No. 2	No. 3	No. 4	No. 5	No. 6	No. 1	No. 2	No. 3	No. 4	No. 5	No. 6
F/M RATIO $\frac{\text{Lb BOD removed/day}}{\text{Lb MLVSS}}$	0.05	0.07	0.12	0.20	0.23	0.37	0.34	0.23	0.16	0.18	0.23	0.18
F/M RATIO $\frac{\text{Lb BOD applied/day}}{\text{Lb MLVSS}}$	0.05	0.07	0.13	0.21	0.25	0.41	0.44	0.27	0.17	0.19	0.24	0.20
OXYGEN UPTAKE $\frac{\text{Lb oxygen}}{\text{Lb MLVSS-Day}}$	0.14	0.11	0.18	0.29	0.52	0.44	0.40	0.28	0.23	0.25	0.31	0.24
1/SRT	0.03	0.06	0.07	0.09	0.09	0.33	0.14	0.08	0.04	0.04	0.02	0.05

Wastewater temperature becomes an important factor in design of activated sludge treatment systems and aerated lagoons, especially in the case of aerated lagoons where significant heat loss can be expected during winter months due primarily to the large surface area of the aeration basin. For temperatures below 55°-66° F, BOD removal efficiency can be expected to decrease significantly.[8]

The effect of temperature on the BOD removal rate is usually expressed by the equation,

$$k_T = k_{20} \; \Theta^{T-20}$$

where k_T = BOD removal rate coefficient at temperature T

k_{20} = BOD removal rate coefficient at 20° C

T = temperature, °C

Θ = temperature coefficient

Pilot plant studies should be conducted under cold weather as well as summer operating conditions if possible. An appropriate temperature correction factor should be considered in design. Values of Θ most commonly range between 1.02 and 1.09, depending on the type of treatment process and operating conditions.

PROBLEM 3.24

An industrial waste will be treated using a complete mix activated sludge process. Based on laboratory and pilot plant studies the design criteria shown in Table 3.14 will be used. Calculate the following:

(a) aeration volume

(b) waste sludge production

(c) final clarifier area

(d) oxygen requirement

(e) aeration horsepower requirement

Table 3.14

DESIGN CRITERIA FOR PROBLEM 3.24

Influent and Effluent Design Criteria

Average daily influent BOD_5, S_o: 2000 mg/l

Maximum daily influent BOD_5: 2200 mg/l

Effluent BOD_5 (total): 200 mg/l

Effluent BOD_5 (soluble): 150 mg/l

Influent suspended solids: 500 mg/l

Effluent suspended solids: 100 mg/l

Average daily flow: 1.4 mgd

Maximum daily flow: 2.0 mgd

Treatment Unit Design Criteria

k: 0.0017 1/mg-day @ 20° C

Y: 0.64

b: 0.012 $days^{-1}$

F/M: 0.15-0.30 lbs BOD applied/lb MLVSS

SRT: minimum 8 days

Summer wastewater temperature: 22° C

Winter wastewater temperature: 10° C

Θ: 1.04

MLVSS: 2500 mg/l

MLSS: 3500 mg/l

Sludge recycle rate: 40%

Clarifier solids loading: 15 $lbs/day\text{-}ft^2$

Clarifier hydraulic loading: 600 gpd/ft^2

Aeration Design Criteria

Y': 0.6 lbs oxygen/lb BOD removed

b': 0.1 lbs oxygen/day per lb MLVSS

α: 0.8

β 0.9

N_o: 3.0 lbs oxygen/hp-hr

C_{sat}: 8.5 mg/l

C_L: 2.0 mg/l

Θ: 1.024

Solution

(a) Calculate aeration basin volume

Summer wastewater temperature = 22° C

Winter wastewater temperature = 10° C

$k = 0.0017$ 1/mg-day @ 20° C

$\Theta = 1.04$

$k_{22°} = 1.04^{22-20}(0.0017) = 0.00184$ 1/mg-day

$k_{10°} = 1.04^{10-20}(0.0017) = 0.00119$ 1/mg-day

Calculate aeration basin volume based on summer conditions with k = 0.00184 1/mg-day.

Solids retention time (SRT) may be calculated from the equation,

$1/SRT = Y\ (F/M) - b$

Using the relationship shown in Figure 3.11, Problem 3.23,

$F/M = k\ S_e$

$F/M = (0.00184\ \text{1/mg-day})(150\ \text{mg/l}) = 0.276\ \text{days}^{-1}$

$1/SRT = 0.64\ (0.276) - 0.012 = 0.16464\ \text{days}^{-1}$

$SRT = 6.1$ days

Based on design criteria a minimum SRT of 8 days is required for design.

Calculate aeration basin volume from the equation (Problem 3.22),

$$XV = \frac{Y\ (S_o - S_e)\ SRT}{1 + b(SRT)}\ Q$$

$$XV = \frac{0.64(2000-150)(8)(1.4)}{1 + 0.012(8)}$$

$XV = 12,100$

For MLVSS = 2500 mg/1

$$V = \frac{12,100}{2500} = 4.84 \text{ million gallons}$$

Calculate aeration volume based on winter conditions with k = 0.00119 1/mg-day.

$$F/M = (0.00119)(150) = 0.179 \text{ days}^{-1}$$

$$1/SRT = 0.64(0.173) - 0.012 = 0.102 \text{ days}^{-1}$$

$$SRT = 9.8 \text{ days or 10 days for design}$$

$$XV = \frac{0.64(2000-150)(10)(1.4)}{1 + 0.012(10)} = 14,800$$

Assume MLVSS of 2500 mg/1 for winter conditions.

$$V = \frac{14,800}{2500} = 5.92 \text{ million gallons}$$

$$\text{Detention time} = \frac{5.92 \text{ mg}}{1.4 \text{ mgd}} = 4.3 \text{ days}$$

Therefore, winter conditions control and design is based on 5.92 million gallon aeration volume.

Check F/M ratio.

$$F/M = \frac{1.4 \text{ mgd} \times 2000 \text{ mg/1} \times 8.34}{5.92 \text{ mg} \times 2500 \text{ mg/1} \times 8.34} = 0.19 \text{ days}^{-1}$$

Check F/M ratio at maximum flow and concentration.

$$F/M = \frac{2.0 \text{ mgd} \times 2200 \text{ mg/1} \times 8.34}{5.92 \text{ mg} \times 2500 \text{ mg/1} \times 8.34} = 0.30 \text{ days}^{-1}$$

Both F/M ratios are within design criteria range.

(b) Calculate waste sludge production

$$SRT = \frac{\text{lbs MLSS}}{\text{lbs solids wasted/day + lbs effluent solids}}$$

Calculate pounds solids wasted/day for a minimum SRT of 8 days. MLSS concentration = 3500 mg/1.

$$SRT = \frac{5.92 \text{ mg} \times 8.34 \times 3500 \text{ mg/l}}{Q_w + (1.4 \text{ mgd} \times 8.34 \times 100 \text{ mg/l})}$$

where Q_w = lbs solids wasted/day

Q_w = 21,400 lbs/day

(c) Calculate final clarifier area

Calculate solids loading to final clarifier based on 40% return sludge rate, R, and MLSS concentration of 3500 mg/l.

(Q + R)(MLSS)(8.34) = solids loading

(1.4 + 0.56)(3500)(8.34) = 57,200 lbs/day

Clarifier area required for thickening is calculated from allowable solids loading of 15 lbs/day-sq. ft.

$$\frac{57{,}200 \text{ lbs/day}}{15 \text{ lbs/day-sq.ft.}} = 3813 \text{ ft}^2$$

Clarifier area based on hydraulic loading is calculated from maximum flow rate and 600 gpd/sq. ft. loading.

$$\frac{2.0 \times 10^6 \text{ gpd}}{600 \text{ gpd/ft}^2} = 3333 \text{ ft}^2$$

Therefore, area required for thickening controls and minimum clarifier area of 3800 square feet is required.

(d) Calculate oxygen requirements

lbs O_2 required/day = Y'(lbs BOD removed/day) + b' (lbs MLVSS)

lbs BOD removed/day = 1.4 mgd x (2000-150) x 8.34
= 21,600 lbs/day

lbs MLVSS = 2500 mg/l x 5.9 mg x 8.34
= 123,000 lbs

$Y' = 0.6$

$b' = 0.1$

$\text{lbs } O_2 = 0.6\ (21{,}600) + 0.1\ (123{,}000)$

$\text{lbs } O_2 = 25{,}300 \text{ lbs/day} = 1050 \text{ lbs/hr}$

(e) Calculate aeration horsepower requirement

Calculate oxygen transfer rate for design conditions from the equation (Problem 3.19),

$$N = N_o \left[\frac{\beta \times (C_{sat} - C_L)}{C_{sc}}\right] 1.024^{T-20} \alpha$$

Aeration requirement is based on summer basin temperature.

$$N = 3.0 \left[\frac{0.9 \times (8.5-2.0)}{9.2}\right] 1.024^{22-20} (0.8)$$

$N = 1.6 \text{ lbs } O_2\text{/hp-hr}$

$$\text{Horsepower} = \frac{1050 \text{ lbs } O_2\text{/hr}}{1.6 \text{ lbs } O_2\text{/hp-hr}} = 655 \text{ hp}$$

Check power level required for mixing. Assume 100 hp/million gallons required.

$$\frac{655 \text{ hp}}{5.92 \text{ mg}} = 110 \text{ hp/million gallons}$$

Therefore, 655 horsepower should be adequate for oxygen transfer and mixing requirements.

Design Summary

Aeration volume = 5.92 million gallons

Detention time = 4.3 days

Waste sludge production = 21,400 lbs/day

Secondary clarifier area = 3800 ft^2

Aeration horsepower = 655 hp

Oxygen requirement = 25,300 lbs/day

Industrial wastes that are low in nutrients will also require addition of nitrogen and/or phosphorus for biological treatment.

References

1. U.S. Environmental Protection Agency, "Process Design Manual for Phosphorus Removal," October, 1971

2. Stepko, W.E. and W.H. Schroeder, "Design Consideration to Attain Less Than 0.3 mg/l Effluent Phosphorus," in High Quality Effluents Seminar - Conference Proceedings No. 3, Ontario Ministry of the Environment, Toronto, Ontario, 1976, p. 181.

3. National Research Council: Trickling Filters (in Sewage Treatment at Military Installations), _Sewage Works Journal_, Vol. 18, No. 5, 1946.

4. Eckenfelder, W. Wesley, "Trickling Filters" in _Process Design in Water Quality Engineering-New Concepts and Developments_, edited by Edward L. Thackston and W.W. Eckenfelder, Jenkins Publishing Company, New York, 1972, pp. 115-120.

5. Fair, Gordon M. and John C. Geyer, _Elements of Water Supply and Waste Disposal_, John Wiley and Sons, New York, 1958, p. 847.

6. Lawrence, Alonzo, W. and Perry L. McCarty, "Unified Basis for Biological Treatment Design and Operation," _Jour. San. Eng. Div., Proc. Amer. Soc. Civil Engrs._, Vol. 96, No. SA3, June, 1970, pp. 757-777.

7. _Ibid._, pp. 757-77.

8. Bartsch, Eric H. and Clifford W. Randall, "Aerated Lagoons - A Report on the State of the Art," _Journal Water Pollution Control Federation_, Vol. 43, No. 4, April, 1971, pp. 699-708.

Additional References

1. Metcalf & Eddy, Inc., Wastewater Engineering, McGraw-Hill, Inc., New York, 1972.

2. Harbold, Harry S., "How to Control Biological Waste Treatment Process," Chemical Engineering, December 6, 1976.

3. Clark, John W., Warren Viessman, Mark J. Hammer, Water Supply and Pollution Control, Harper and Row Publishers, Inc., New York, 1977.

4. Eckenfelder, W. Wesley, "Comparative Biological Waste Treatment Design," Jour. San. Eng. Div., Proc. Amer. Soc. Civil Engrs., Vol. 93, No. SA6, December, 1967, pp. 157-170.

Chapter 4

SLUDGE TREATMENT AND DISPOSAL

PROBLEM 4.1

Sludge processing prior to final disposal may be divided into the following unit operations: thickening, stabilization, conditioning, dewatering, and thermal reduction.

(a) Identify several processes commonly associated with each unit operation.

(b) Identify the type of sludge or process most applicable to each of the sludge thickening alternatives identified in (a).

(c) List the relative advantages and disadvantages of anaerobic and aerobic digestion.

Solution

(a) Thickening

Gravity Thickening
Dissolved Air Flotation
Centrifuge

Stabilization

Anaerobic Digestion
Aerobic Digestion
Lime Treatment
Composting

Conditioning

Chemical Conditioning

Conditioning

Heat Treatment

Dewatering

Rotary Vacuum Filter
Centrifuge
Belt Filter Press
Pressure Filter
Drying Beds
Lagoons

Thermal Reduction

Incineration
Wet Air Oxidation

(b) Gravity thickening is generally best suited for treatment of primary sludge or combined primary and trickling filter sludge. Gravity thickening may be applicable to activated sludge or combined primary and activated sludge. Polymer or other flocculants are generally used in such cases to improve thickening.

Dissolved air flotation is generally used for waste activated sludge. Polymers are often used to improve solids capture and degree of solids concentration.

Centrifuges are generally used for waste activated sludge or industrial sludges that are difficult to dewater.

(c) Anaerobic Digestion

Advantages	Disadvantages
Methane gas production	Requires close process control
High organic loadings possible	External heat required
Low solids production	Supernatant requires treatment
Reduction of volatile organic matter	High capital cost

Aerobic Digestion

Advantages	Disadvantages
Supplemental heating usually not required	Higher operating cost
Relatively easy to operate	
Does not generate significant odors	Requires energy input for aeration
Supernatant of relatively good quality	

Aerobic digestion is generally used to treat only waste activated sludge, and has been used in many smaller plants because of the simplicity of operation.

PROBLEM 4.2

Calculate the waste sludge production for the treatment plant design criteria given in Problem 3.2. Assume the raw sludge contains 97.5% water and 75% volatile solids, and is to be treated by anaerobic digestion. The following design criteria are applicable for the digester: 60% volatile solids destruction, 30 day detention time, raw sludge temperature = 21° C, digested sludge solids = 5.0%. Calculate the amount of digested sludge produced daily, gas production, and digester volume required.

Solution

(a) Calculate sludge production from primary clarifier. Assume 60% total suspended solids removal.

0.2 lbs/cap-day x 32,000 = 6,400 lbs/day

6400 x 0.6 = 3,840 lbs primary sludge/day

Calculate waste activated sludge production.

Assume 0.55 lbs solids/lb BOD removed.

Lbs BOD to aeration = 3700 lbs/day

Effluent BOD = 30 mg/l x 8.34 x 3.2 mgd = 800 lbs/day

(3700 - 800) x 0.55 = 1595 lbs sludge/day

Total sludge to digester = 5435 lbs/day dry solids

(b) Assume sludge specific gravity = 1.0

$$\text{Volume sludge} = \frac{5435 \text{ lbs/day}}{0.025} = 217{,}000 \text{ lbs/day}$$

= 26,020 gallons/day

= 3480 ft^3/day

Volatile solids = 5435 x 0.75 = 4100 lbs/day to digester

Non-volatile solids = 5435 - 4100 = 1335 lbs/day to digester

Volatile solids = 4100 x 0.6 = 2460 lbs/day destroyed

Volatile solids = 4100 - 2460 = 1640 lbs/day out

Assume that pounds non-volatile solids in raw sludge is equal to pounds non-volatile solids in digested sludge. For digested sludge,

Non-volatile solids = 1335 lbs/day

Volatile solids = 1640 lbs/day

$$\text{Volume sludge out of digester} = \frac{1335 + 1640}{0.05} = 59{,}500 \text{ lbs/day}$$

= 7100 gallons/day

(c) Calculate gas production. Assume gas is 72% methane and 28% carbon dioxide by volume. Assume 8.5 cubic feet of gas is produced per pound of volatile solids to digester.

Gas production = 8.5 ft^3/lb x 4100 lbs/day

= 34,900 ft^3/day

Methane production = 34,900 x 0.72 = 25,100 ft^3/day

(d) Calculate digester volume for hydraulic detention time of 30 days and sludge flow rate of 3480 cubic feet/day.

Volume = 30 x 3480 = 104,400 ft^3

Assume a 20 foot side water depth and cylindrical shaped digester.

Use 4 digesters, each 41 foot diameter x 20 foot side water depth.

Total volume = 105,600 cubic feet.

PROBLEM 4.3

For Problem 4.2, calculate if enough heat is present to raise the temperature of the incoming sludge to 35° C if the average ambient winter temperature is 7° C and summer temperature is 19° C. Assume an overall heat transfer coefficient for the digester of 0.14 BTU/hr-sq. ft.-°F. Raw sludge temp. = 21°C

Solution

Digester heat requirement = heat required to raise digester temperature to 35° C + digester heat transfer losses.

Calculate heat required to raise sludge from 21° C to 35° C.

$$217{,}000\ \frac{\text{lbs}}{\text{day}} \times 1.0\ \frac{\text{BTU}}{\text{lb} - {}^\circ\text{F}} \times (95.0^\circ\ \text{F} - 69.8^\circ\ \text{F}) = 5.4 \times 10^6\ \frac{\text{BTU}}{\text{day}}$$

Calculate digester heat transfer losses using heat transfer equation,

$$Q = U\ A\ \Delta T$$

where Q = heat transfer rate, BTU/hr

U = overall heat transfer coefficient

ΔT = temperature difference, °F

A = heat transfer area, ft^2

Surface area/digester = roof area + floor area + side area

$$= \frac{\pi(41)^2}{4} + \frac{\pi(41)^2}{4} + \pi(41)(20)$$

$$= 5214\ ft^2/\text{digester}$$

$$\text{Total area} = 20{,}860\ ft^2$$

For summer conditions,

$$Q = 0.14\ \frac{BTU}{hr\text{-}ft^2\text{-}°F} \times 20{,}860\ ft^2 \times (95.0°\ F - 66.2°\ F)$$

$$Q = 84{,}100\ BTU/hr = 2.0 \times 10^6\ BTU/day$$

For winter conditions,

$$Q = 0.14 \times 20{,}860 \times (95.0 - 44.6)$$

$$Q = 147{,}200\ BTU/hr = 3.5 \times 10^6\ BTU/day$$

Total heat required (winter) $= 5.4 \times 10^6 + 3.5 \times 10^6$
$= 8.9 \times 10^6$ BTU/day

Total heat required (summer) $= 5.4 \times 10^6 + 2.0 \times 10^6$
$= 7.4 \times 10^6$ BTU/day

Calculate heat produced from methane. The heating value of methane is 963 BTU/cu. ft. From Problem 4.2, approximately 25,000 cu. ft./day methane is produced.

$$25{,}000\ \frac{ft^3}{day} \times 963\ \frac{BTU}{ft^3} = 24 \times 10^6\ BTU/day$$

Assuming a combustion efficiency of 60%,

$$24 \times 10^6\ \frac{BTU}{day} \times 0.6 = 14.4 \times 10^6\ \frac{BTU}{day}\ \text{available for heating}$$

Based on these calculations, 62% of available methane would satisfy winter heating requirements and 51% methane would satisfy summer heating requirements.

PROBLEM 4.4

Assume that for a 15 mgd wastewater treatment plant 35,000 pounds/day of dry solids are treated in an anaerobic digester. The solids are 73% volatile and 65% of the volatile solids are destroyed. Fourteen standard cubic feet of methane gas is produced for each pound of volatile solids destroyed. It is proposed to store the gas at 40 psi and use the gas for power generation in the plant for an internal combustion engine. Daily power requirement at the plant is 416 kilowatts. Determine the horsepower available from an internal combustion engine and the electrical power generated. Assume an engine efficiency of 35% and a generator efficiency of 30%.

Solution

Volatile solids destroyed = 35,000 x 0.73 x 0.65
= 16,600 lbs/day

Gas produced = 16,600 x 14 = 232,400 scf/day

Assume gas is 70% methane. Heating value for methane is 963 BTU/scf.

Heat available = 232,400 x 963 x 0.70
= 156×10^6 BTU/day

Calculate power available. Use 2545 BTU/hp-hr at 35% efficiency or 7270 BTU/hp-hr.

$$\frac{156 \times 10^6 \text{ BTU/day}}{7270 \text{ BTU/hp-hr} \times 24} = 894 \text{ hp/day}$$

Calculate electrical power generated at 30% efficiency.

$$\frac{2545 \text{ BTU/hp-hr}}{0.7457 \text{ kwh/hp-hr} \times 0.30} = 11{,}375 \text{ BTU/kwh}$$

$$\frac{156 \times 10^6}{11{,}375 \times 24} = 570 \text{ kw/day}$$

Therefore, enough power could be generated to satisfy the 416 killowatt power requirement for the plant.

PROBLEM 4.5

Waste sludge from a lime treatment process is to be recalcined in a kiln to regenerate lime. The sludge fed to the kiln and the exit flue gas have a composition as shown in Table 1. The fuel gas used is methane at a rate of 4,500 cubic feet/hour (measured at 60° F, 1 atmosphere, dry). Assume 92% of calcium carbonate entering the kiln is converted to quicklime.

(a) Determine the amount of sludge recalcined in tons/day.

(b) Determine the production of lime in tons/day.

TABLE 4.1

Sludge and Flue Gas Analysis. Problem 4.5.

Sludge (Wt. %)		Flue Gas (Vol. %)	
$CaCO_3$	50.2%	CO_2	19.0%
Water	45.7%	CO	0.3%
Inerts	4.1%	N_2	78.9%
		O_2	1.8%

Solution

In recalcination of waste lime sludge the sludge is burned at high temperatures to decompose calcium carbonate to calcium oxide (lime) for reuse in wastewater treatment. A flue gas is produced from the combustion process.

Assume that 100 lb-moles of dry flue gas are produced and make a material balance. For dry flue gas,

$$CO_2 = 19.0 \text{ lb-moles}$$

$$CO = 0.3 \text{ lb-moles}$$

$$N_2 = 78.9 \text{ lb-moles}$$

$$O_2 = 1.8 \text{ lb-moles}$$

Material balance using oxygen. Nitrogen is inert gas and does not react. Air is 79% nitrogen and 21% oxygen by volume.

Entering air volume $= \dfrac{78.9}{0.79} = 99.9$ lb-moles

Oxygen in air $= 99.9 \times 0.21 = 21.0$ lb-moles

Oxygen in flue gas $= 1.8$ lb-moles

Net oxygen demand $= 21.0 - 1.8 = 19.2$ lb-moles

Calculate oxygen required for carbon monoxide combustion.

$$CO + \tfrac{1}{2}\, O_2 = CO_2$$

$$O_2 = \tfrac{1}{2} \times 0.3 = 0.15 \text{ lb-moles}$$

Theoretical oxygen required $= 0.15 + 19.2 = 19.35$ lb-moles.

Calculate theoretical oxygen required for combustion of methane.

$$CH_4 + 2\, O_2 = CO_2 + 2\, H_2O$$

$$O_2 = 2 \times \frac{4500 \text{ ft}^3 \text{ } CH_4}{359 \text{ ft}^3/\text{lb-mole}} = 25 \text{ lb-moles}$$

Actual moles dry flue gas $= \dfrac{25 \text{ lb-moles } O_2}{19.35 \text{ lb-moles } O_2/100 \text{ mole flue gas}}$

$= 129.6$ lb-moles/hr

In recalcination of waste sludge $CaCO_3$ is converted to quicklime (CaO) according to the following equation.

$$CaCO_3 = CaO + CO_2 \text{ (gas)}$$

CO_2 in flue gas $= 129.6 \times 0.19 = 24.6$ lb-moles

CO_2 from methane $= \dfrac{4500 \text{ ft}^3}{359 \text{ ft}^3/\text{lb-mole}} = 12.5$ lb-moles

CO_2 from $CaCO_3$ reaction $= 24.6 - 12.5$

$= 12.1$ lb-moles/hr

$CaCO_3$ weighs 100 lb/lb-mole. Equation for conversion of $CaCO_3$ indicates that 12.1 lb-moles of $CaCO_3$ would be converted to 12.1 lb-moles of CO_2 if conversion were 100% complete. Problem states that 92% $CaCO_3$ is converted.

$$CaCO_3 \text{ in sludge} = \frac{12.1 \text{ lb-moles} \times 100 \text{ lb/mole}}{0.92}$$

$$= 1315 \text{ lbs/hr}$$

(a) From Table 1, sludge is 50.2% $CaCO_3$.

$$\text{Weight sludge} = \frac{1315}{0.502} = 2620 \text{ lbs/hr}$$

$$= 31.4 \text{ tons/day}$$

(b) CaO weighs 56 lbs/mole.

$$\text{Lime produced} = 12.1 \text{ lb-moles} \times 56 \text{ lb/mole}$$

$$= 678 \text{ lbs/hr} = 8.1 \text{ tons/day}$$

PROBLEM 4.6

Sludge containing 25% total solids and 75% volatile solids will be fed to a fluidized bed incinerator at the rate of 1800 pounds/hour (dry basis). The incinerator will operate 8 hours/day, 5 days/week. Operating temperature will be approximately 1400° F. The sludge analysis, on a dry weight basis, is as follows: 43% carbon; 6.37% hydrogen; 0.24% sulfur; 14% ash; 33.35% oxygen. Calculate the following:

(a) Furnace diameter.

(b) Auxiliary fuel requirement (natural gas).

(c) Air supply rate.

(d) Amount of ash produced.

Solution

The following equations have been proposed for fluidized bed incinerator design[1]:

$$B = 93\ P \qquad (1)$$

$$B = 145\ C + 620\ (H - O_x/7.94) + 45S \qquad (2)$$

where B = sludge heat value, BTU/dry lb

P = sludge volatile content, percent

C, H, O_x, S = percentages of carbon, hydrogen, oxygen, and sulfur in sludge

$$\log S_L = 2.7 - 0.0222\ M \qquad (3)$$

$$\log B_r = 5.947 - 0.0096\ M \qquad (4)$$

where S_L = sludge loading rate, dry lb/hr - ft^2

M = sludge moisture content, weight percent

B_r = burning rate, BTU/hr - ft^2

(a) Calculate furnace diameter using equation (3). Sludge contains 75% water. Equation (3) is applicable for furnace temperatures exceeding 1200° F.

$$\log S_L = 2.7 - 0.0222\ (75)$$

$$S_L = 10.8\ \text{lbs/hr} - ft^2$$

$$\text{Area} = \frac{1800\ \text{lbs/hr}}{10.8\ \text{lbs/hr} - ft^2} = 167\ ft^2$$

diameter = 14.6 feet

(b) Auxiliary fuel requirement can be calculated from the difference in the total heat rate and the sludge heat input.

Since sludge analysis is given, calculate sludge heat value from equation (2).

$$B = 145(43) + 620[6.37 - (33.35/7.94)] + 45(0.24)$$

$$B = 7590\ \text{BTU/lb}$$

Sludge heat input = 7590 BTU/lb x 1800 lb/hr

$$= 13.66 \times 10^6 \text{ BTU/hr}$$

Calculate burning rate from equation (4).

$$\log B_r = 5.947 - 0.0096\ (75)$$

$$B_r = 169{,}000 \text{ BTU/hr} - \text{ft}^2$$

$$\text{Total heat rate} = 169{,}000 \text{ BTU/hr} - \text{ft}^2 \times 167 \text{ ft}^2$$

$$= 28.22 \times 10^6 \text{ BTU/hr}$$

$$\text{Auxiliary heat required} = 28.22 \times 10^6 - 13.66 \times 10^6$$

$$= 14.56 \times 10^6 \text{ BTU/hr}$$

Assume natural gas has a heating value of 10,000 BTU/cu. ft. and calculate auxiliary fuel requirement.

$$\frac{14.56 \times 10^6 \text{ BTU/hr}}{10 \times 10^3 \text{ BTU/ft}^3} = 1456 \text{ ft}^3\text{/hr} = 25 \text{ ft}^3\text{/min}$$

(c) Assume 40% excess air is supplied. Theoretical air requirement is based on the following combustion equations, with air 23% by weight oxygen.

$C + O_2 = CO_2$ (11.5 lbs air/lb carbon)

$2H_2 + O_2 = 2H_2O$ (34.3 lbs air/lb hydrogen)

$S + O_2 = SO_2$ (4.3 lbs air/lb sulfur)

Carbon:	11.5 lb/lb x 0.43 x 1800 lb/hr	= 8901 lb/hr
Hydrogen:	34.3 lb/lb x 0.0637 x 1800 lb/hr	= 3933 lb/hr
Sulfur:	4.3 lb/lb x 0.0024 x 1800 lb/hr	= 16 lb/hr
		12,850 lb/hr
	40% excess	5,140 lb/hr
	Total	17,990 lb/hr

Air weighs 0.0749 lb/ft^3 at 70° F.

$$\text{Air flow rate} = \frac{17{,}990 \text{ lb/hr}}{0.0749 \times 60} = 4000 \text{ scfm}$$

(d) Assume ash input = ash output

Ash = 1800 lb/hr x 0.14 = 252 lb/hr

Additional important design considerations include air pollution control equipment, sludge feed equipment, and ash disposal. These should be considered as part of a final design proposal.

PROBLEM 4.7

Consider the following equation[2] for vacuum filtration of a waste activated sludge conditioned with 6 - 19% by weight of ferric chloride.

$$L = 3.85 \frac{C^{0.767}}{T^{0.656}}$$

where L = cake yield, lbs/hr-ft^2

T = cycle time, minutes/revolution

C = percent solids in feed, %

Calculate the required surface area of a vacuum filter operating 16 hours/day for a thickened sludge with a solids concentration of 2.5%. The flow rate of the feed sludge is 52,000 pounds/day on a dry solids basis. Assume a cycle time of 5 minutes and a ferric chloride addition of 6%.

Solution

Calculate filter yield of dry solids for given design conditions,

$$L = 3.85 \frac{(2.5)^{0.767}}{(5)^{0.656}} = 2.7 \text{ lbs/hr-ft}^2$$

Calculate surface area for 16 hour operating day. Assume a

safety factor of 0.8 for design.

$$\frac{52{,}000 \text{ lbs/day}}{16 \text{ hours/day}} = 3250 \text{ lbs/hour}$$

$$\text{Area} = \frac{3250 \text{ lbs/hr}}{2.7 \text{ lbs/hr - ft}^2 \times 0.8} = 1500 \text{ ft}^2$$

PROBLEM 4.8

A dissolved air flotation pilot plant unit having a surface area of 15 square feet is operated at varying air/solids ratio and solids loading rates. Polymer is also added to improve solids capture. Results indicate that a float solids of 4% can be achieved for the following operating conditions:

Polymer dosage = 3.5 lbs/ton

Air/solids ratio = 0.02

Solids loading = 55 lbs/hr

Calculate the size unit required for a sludge flow rate of 220,000 gallons/day and a solids concentration of 1.0%. Assume the unit is to operate on a 35 hour/week schedule.

Solution

$$\text{Dry solids} = 220{,}000 \text{ gal/day} \times 8.34 \text{ lbs/gal} \times 0.01$$
$$= 18{,}348 \text{ lbs/day}$$

$$\text{Solids loading} = \frac{18{,}348 \text{ lbs/day} \times 7 \text{ days/week}}{35 \text{ hours/week}}$$
$$= 3670 \text{ lbs/hr}$$

Calculate solids loading rate based on pilot plant operation.

$$\frac{55 \text{ lbs/hr}}{15 \text{ ft}^2} = 3.67 \text{ lbs/hr-ft}^2$$

$$\text{Total area} = \frac{3670 \text{ lbs/hr}}{3.67 \text{ lbs/hr - ft}^2}$$
$$= 1000 \text{ ft}^2$$

Select two units, each a minimum of 500 square feet.

Check hydraulic loading rate.

$$\frac{220{,}000 \text{ gal/day x 7 days/week}}{35 \text{ hrs/week x 60 x 1000 ft}^2} = 0.73 \text{ gpm/ft}^2$$

PROBLEM 4.9

A wastewater sludge with the following characteristics is to be dewatered using a filter press.

Dry solids concentration	6%
Specific gravity	1.02
Dry solids density	187 lbs/ft^3

The sludge flow rate is 100 gallons/minute. Based on pilot plant results, a cake solids of 30% can be achieved with a total cycle time of 120 minutes. Calculate the following:

(a) Density of the cake solids.

(b) Volume of the dewatered cake.

(c) Filter press volume based on operation of 24 hours/day, 7 days/week.

Solution

(a) Calculate density of the solids cake, C_d,

$$\frac{C_d}{62.4 \text{ lb/ft}^3} = \frac{187 \text{ lb/ft}^3}{187 \text{ lb/ft}^3 - 0.3\ (187 \text{ lb/ft}^3 - 62.4 \text{ lb/ft}^3)}$$

$$C_d = 77.9 \text{ lb/ft}^3$$

(b) Calculate volume of dewatered cake.

$$\text{Sludge flow} = 100\ \frac{\text{gal}}{\text{min}} \text{ x } 1440\ \frac{\text{min}}{\text{day}} \text{ x } 8.34\ \frac{\text{lb}}{\text{gal}} \text{ x } 1.02$$

$$= 1.22 \text{ x } 10^6 \text{ lbs/day}$$

Dry solids flow = $(1.22 \times 10^6) \times 0.06$ = 73,500 lbs/day

$$\text{Volume dewatered cake} = \frac{73{,}500 \text{ lbs/day}}{0.3 \times 77.9 \text{ lbs/ft}^3} = 3145 \text{ ft}^3\text{/day}$$

(c) Calculate filter press volume.

$$\text{Cycles/day} = \frac{24 \text{ hrs}}{2 \text{ hrs/cycle}} = 12 \text{ cycles/day}$$

$$\text{Volume} = \frac{3145 \text{ ft}^3\text{/day}}{12 \text{ cycles/day}} = 262 \text{ ft}^3$$

PROBLEM 4.10

It has been recommended that to optimize anaerobic digester performance a bicarbonate alkalinity concentration of 3500 mg/l to 5000 mg/l be maintained for buffering.[3] Assume that the plant operator measures the total alkalinity and volatile acids concentrations for the digester on a routine basis.

(a) Calculate the bicarbonate alkalinity if total alkalinity is 5900 mg/l (as $CaCO_3$) and volatile acids is 2430 mg/l (as acetic acid).

(b) Determine the amount of bicarbonate alkalinity to be added to a digester with an alkalinity of 2000 mg/l necessary to achieve an alkalinity of 4000 mg/l. Digester detention time is 20 days.

Solution

(a) Bicarbonate alkalinity may be calculated from the equation,

Bicarbonate alk. = total alk. - (0.71) (volatile acids)

Bicarbonate alkalinity = 5900 - (0.71) (2430)

= 4175 mg/l as $CaCO_3$

Bicarbonate and total alkalinity are measured as mg/l $CaCO_3$, volatile acids as mg/l acetic acid in this equation.

(b) The dosage of bicarbonate alkalinity required can be calculated from the equation.[4]

$$D_d = D_{max} (1 - e^{-\lambda})$$

where D_d = daily dose of alkalinity necessary to reach required level, mg/l as $CaCO_3$

D_{max} = required increase in alkalinity concentration, mg/l

λ = reciprocal of average digester detention time, $days^{-1}$

$$D_d = (4000 - 2000)(1 - e^{-0.05})$$

$$D_d = 100 \text{ mg/l as } CaCO_3$$

PROBLEM 4.11

Assume that anaerobic digester performance will be controlled by maintaining the pH level at 7.0. The carbon dioxide content of the digester gas is routinely measured. Calculate the theoretical bicarbonate alkalinity concentration required when the gas contains 34% CO_2 by volume.

Solution

The equilibrium equation for the system may be written as,

$$H_2CO_3 = HCO_3^- + H^+$$

H_2CO_3 refers to the sum of dissolved CO_2 and H_2CO_3.

The CO_2 concentration is assumed equal to the H_2CO_3 concentration.

$$K_1 = \frac{[H^+][HCO_3^-]}{[CO_2]}$$

$$pH = pK_1 + \log \frac{[HCO_3^-]}{[CO_2]}$$

$$pK_1 = 6.35$$

In a closed system with 100% CO_2 gas at 1 atmosphere pressure, the CO_2 concentration is approximately 1320 mg/l.

$$34\%\ CO_2 = 449\ \text{mg/l}$$

$$\frac{.449\ \text{g/l}}{44\ \text{g/mole}} = 0.0102\ \text{moles/liter}$$

$$7.0 = 6.35 + \log\ [HCO_3^-/0.0102]$$

$$[HCO_3^-] = 0.0456\ \text{moles/liter}$$

Bicarbonate alkalinity = 0.0456 moles/liter x 61 g/mole
= 2782 mg/l as $CaCO_3$

PROBLEM 4.12

Land application of sewage sludge is to be evaluated as a sludge disposal alternative. Metals in sludge are of environmental concern due to potential long term reductions in soil productivity and possible accumulation in the human food chain. Assume that a regulatory agency requires that the maximum rate of application of heavy metals to agricultural land not exceed the following levels during the period of time in which the land is used:

Metal	Loading (lbs/acre)
Cadmium	10
Copper	250
Nickel	100
Lead	1000
Zinc	1000

(a) Table 4.2 shows the analysis to two sludges being considered for land application. Sludge A is from a treatment plant treating domestic wastewater. Sludge B is from a plant receiving domestic and industrial wastewater. Calculate the allowable application rate for each sludge in tons/acre based on the above guidelines for heavy metals.

(b) Discuss the significance of soil cation exchange capacity and pH relative to land application of sludge.

(c) Identify those crop species generally considered more tolerant to heavy metal toxicity.

(d) Discuss the significance of cadmium, zinc, copper, and nickel with regard to their effect on plant growth and human toxicity.

TABLE 4.2

Sludge Analysis for Problem 4.12.

	Concentration (mg/kg)	
Metal	Sludge A	Sludge B
Cadmium	20	655
Copper	750	1540
Nickel	110	450
Lead	239	1500
Zinc	1610	5200

Solution

(a) Concentrations of heavy metals are generally expressed on a dry weight basis as mg/kg. For example, consider Sludge A that contains 20 mg/kg cadmium.

$$\frac{20 \text{ mg Cd}}{\text{kg sludge}} = \frac{20 \text{ lbs Cd}}{10^6 \text{ lbs sludge}}$$

Allowable cadmium loading = 10 lbs/acre
The total amount of sludge allowed based on cadmium is calculated as,

TABLE 4.3

Total Amount of Sludge to be Applied Based on Heavy Metal Loadings. Problem 4.12.

Metal	Allowable Loading (tons/acre)	
	Sludge A	Sludge B
Cadmium	250	7.6
Copper	167	81
Nickel	455	110
Lead	2093	333
Zinc	311	96

$$\frac{10^6 \text{ lb sludge}}{20 \text{ lb Cd}} \times \frac{10 \text{ lbs Cd}}{\text{acre}} \times \frac{1 \text{ ton}}{2000 \text{ lbs}} = 250 \text{ tons/acre}$$

Similar calculations are made for each metal and results are tabulated in Table 4.3. Based on these results, the total sludge loading for Sludge A would be limited to 167 tons/acre due to the copper in the sludge. For Sludge B, the total sludge loading could only be 7.6 tons/acre due to the cadmium in the sludge. It is important to note that this is the total sludge application rate over the life of the project. For example, if the land is to be used for 5 years, the sludge loading rate on an annual basis for Sludge A would be 33.4 tons/acre-year.

Many other environmental and social factors should also be considered when evaluating land application as a sludge disposal alternative.

(b) The cation exchange capacity of the soil, measured in meq/100 grams, is a measure of the binding property of all cations including metal cations. A soil with a high cation exchange capacity is more acceptable for land application of sludge than a soil with a lower cation exchange capacity.

Soil pH is an important parameter that affects heavy metal uptake by plants grown in a sludge amended soil. The minimum pH should be maintained in the range of 6.5 - 7.0 to reduce the availability of heavy metals, and can be maintained in this range by lime addition.

(c) Different types of grasses, such as perennial ryegrass and fescue, are most tolerant to metal toxicity. Various farm crops such as corn and grains are moderately tolerant. Many vegetables such as beets, tomato, beans, are most sensitive to metal toxicity.

(d) Zinc, copper, and nickel are most likely to be toxic to plants. Cadmium may be considered toxic to man, while zinc is toxic to plants. Zinc appears to compete with cadmium for plant uptake and the cadmium/zinc ratio is often discussed. A sufficiently low cadmium/zinc ratio protects against excessive cadmium intake by man or animals eating crops from sludge amended soils, and will result in zinc toxicity damage to crops before the cadmium level becomes dangerously high.

PROBLEM 4.13

The raw sludge from a primary treatment plant with a flow rate of 0.6 mgd is analyzed as follows: 5% solids, 30% fixed solids, 70% volatile solids. Specific gravity of organic and inorganic solids are 1.1 and 2.2, respectively.

(a) Calculate the specific gravity of the sludge.

(b) Determine the volume of sludge collected on a daily basis if influent suspended solids concentration is 175 mg/l and effluent suspended solids concentration is 100 mg/l.

Solution

(a) Calculate the specific gravity of the sludge solids from the equation,

$$\frac{1}{S_s} = \frac{0.30}{S_f} + \frac{.70}{S_v}$$

where S_s = specific gravity of sludge solids

S_f = specific gravity of fixed solids (inorganic)

S_v = specific gravity of volatile solids (organic)

$$\frac{1}{S_s} = \frac{0.30}{2.2} + \frac{0.70}{1.1}$$

$$S_s = 1.29$$

Calculate specific gravity of the sludge with 95% water content.

$$\frac{1}{S} = \frac{0.05}{S_s} + \frac{0.95}{S_w}$$

where S = specific gravity of sludge

S_w = specific gravity of water (1.0 assumed)

$$\frac{1}{S} = \frac{0.05}{1.29} + \frac{0.95}{1.0}$$

$$S = 1.01$$

(b) 0.6 mgd x 8.34 x (175 - 100) = 375.3 lbs/day dry solids removed in plant

$$\text{Total lbs sludge} = \frac{375.3}{.05} = 7506 \text{ lbs/day}$$

7506 - 375 = 7131 lbs/day water

Calculate sludge volume

$$\left[\frac{375}{1.29} + \frac{7131}{1.0}\right] \frac{ft^3}{62.4 \text{ lbs}} \times \frac{7.48 \text{ gal}}{ft^3} = 890 \text{ gal/day}$$

References

1. Liao, Paul B., "Fluidized-Bed Sludge Incinerator Design," J. Water Pollution Control Federation, Vol. 46, No. 8, August 1974, pp. 1895-1913.

2. U.S. Environmental Protection Agency, "Continued Evaluation of Oxygen Use in Conventional Activated Sludge Processing," Report No. 17050 DNW 02/72, February, 1972.

3. Brovko, N. and K.Y. Chen, "Optimizing Gas Production, Methane Content, and Buffer Capacity in Digester Operation," Water and Sewage Works, Vol. 124, No. 7, July 1977, pp. 54-57.

4. Barber, Nicholas R. and Carl W. Dale, "Increasing Sludge-Digester Efficiency," Chemical Engineering, July 17, 1978, p. 149.

Additional References

1. Campbell, H.W., R.J. Rush and R. Tew, Sludge Dewatering Design Manual, Research Report No. 72, Ontario Ministry of the Environment, Toronto, Ontario.

Chapter 5

SANITARY ENGINEERING ANALYSIS

PROBLEM 5.1

An existing secondary treatment plant located in the eastern United States treats an average daily flow rate of 250,000 gallons per day. The facility is required to upgrade to meet an effluent limitation of 10 mg/l BOD_5 and 10 mg/l suspended solids. The possibility of using existing facilities followed by spray irrigation is to be investigated. Based on a preliminary survey, four possible spray irrigation sites shown in Table 5.1 will be evaluated.

(a) Describe a possible treatment scheme using spray irrigation.

(b) Determine the land area required for the spray irrigation system.

(c) Describe briefly the feasibility of using each site for spray irrigation based on the information provided.

Solution

(a) A possible land application treatment scheme would include existing facilities, disinfection, holding pond, wastewater pumping and distribution capabilities, and spray irrigation site. A holding pond with a maximum 60 to 90 day detention time may be recommended in the

Table 5.1

Potential Site No.	Land Area (acres)	Pumping Distance (feet)	Pumping Head (feet)	Dwellings Near Site	Average Ground Surface Slope
1	150	1,000	120	Several	5-10%
2	175	8,000	175	Few	5%
3	200	15,000	110	Few	8-10%
4	35	3,500	210	Very Few	10%

Potential Site No.	Type of Ground Cover	Existing Drainage	Soil Permeability
1	Woodland	Fair to Poor	Slow to Moderately Slow
2	Open Field	Good to Moderate	Moderately Rapid
3	Orchard and Open Field	Good to Moderate	Moderately Slow to Moderate
4	Open Field, Agriculture Use	Good	Moderate to Moderately Rapid

eastern United States due to freezing temperatures in the winter months that prevent land application.

(b) Determine required land area from the equation,[1]

$$F = \frac{36.8\ Q}{L\ R}$$

where F = field area, acres

Q = annual flow rate, 10^6 gallons/year

L = period of application, weeks/year

R = rate of application, inches/week

$Q = 0.25 \text{ mgd} \times 365 = 91.25 \times 10^6$ gal/year

Use an application period of 45 week/year. Assume an application rate of 1.5 to 2.0 inches/week will be used for a slow rate system. A more detailed soil analysis and water balance should be made to determine the optimum hydraulic loading rate.

For a 1.5 inch/week application rate,

$$F = \frac{36.8 \times 91.25}{45 \times 1.5} = 50 \text{ acres}$$

Add 15% buffer = 58 acres

For a 2.0 inch/week application rate,

F = 37 acres = 43 acres with 15% buffer

State and local requirements may affect land area. The state of Pennsylvania, for example, recommends that spray irrigation normally be carried out only one day a week on a given section of land, allowing six days for the soil to dry out and reaerate.[2]

(c) Factors to be considered based on information provided include land area and use, location, ownership, soil characteristics, and capital and operating costs.

Site 4 is eliminated since not enough land area is available.

Site 1 is a wooded area that would provide some protection to the spray area during periods of wind and extreme cold. However, the land is apparently near a heavy residential area, possibly making the site undesirable or impossible to use. Existing drainage and soil permeability appear to be less than satisfactory. All sites have a ground slope less than 10% which is generally acceptable. Site 1 is also located closest to the treatment plant, reducing pumping costs.

Site 2 is open farm land and in the eastern United States would be subject to freezing temperatures. A larger holding pond than for a wooded area may be required. Soil conditions appear satisfactory.

Site 3 is located the longest distance from the treatment plant, increasing wastewater conveyance costs. The soil would appear to be well drained. Unlike sites 1 and 2 that may be assumed to be city owned land, site 3 is

Table 5.2

Background Information Needs and Probable Sources for Land Application of Wastewater

Item		Information Source
SOILS:	depth drainage water table classification (mapping) infiltration (permeability) surface slope general properties	County Gov't. U.S. Soil Conservation Service State University Extension Service
WATER:	groundwater yield groundwater elevation and contour groundwater aquifers flood plain streamflow water table depth	State Environmental Protection Agency U.S. Geological Survey U.S. Corps of Engineers
CLIMATE:	temperature rainfall	U.S. Weather Bureau
AGRICULTURE:	land uses crops irrigation land application guidelines	U.S. Soil Conservation Service County agriculture extension service Local and County Gov't. State University extension service U.S. Dept. of Agriculture State Environmental Protection Agency
GEOLOGY:	bedrock type bedrock structure depth to bedrock permeability	State geological survey U.S. Geological Survey U.S. Soil Conservation Service State Environmental Protection Agency

probably privately owned. This may cause problems in acquiring the land and result in significantly higher costs if land has to be purchased or leased.

Prior to making a final decision, several additional factors are important to evaluate in detail. A thorough subsurface soil and rock investigation is necessary. An economic analysis is important. Ground water use as a drinking water supply should be evaluated; monitoring wells are recommended. Other factors such as crop selection, regional site characteristics, public health considerations, hydrology, and social aspects of land use should also be considered. Table 5.2 summarizes information needs and sources for land applications.

PROBLEM 5.2

You have been requested to evaluate several design alternatives for ammonia nitrogen removal.

(a) List at least three treatment alternatives and the relative advantages and disadvantages of each. What considerations would affect your design recommendations?

(b) Why might a high level of ammonia nitrogen in the plant effluent be undesirable?

Solution

(a) Nitrification can be achieved using either a biological or a physical-chemical treatment process. Two biological treatment methods are single stage and two stage biological treatment. Single stage biological treatment combines carbonaceous removal and nitrification in a single step process as shown in Figure 5.1 (a). The two stage process combines a first stage high rate system for carbonaceous removal and a second stage biological process for nitrification as shown in Figure 5.1 (b). Single and two stage biological treatment have the following advantages and disadvantages:

Advantages

Low effluent ammonia levels.
Amenable to modification of existing systems.
Less costly than physical-chemical system.

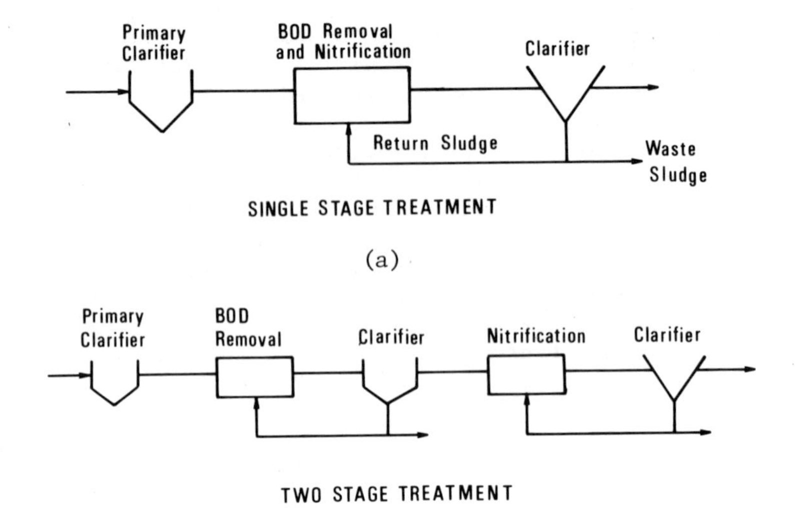

Figure 5.1 (a) Single stage biological nitrification.
(b) Two stage biological nitrification

The single stage treatment system offers the advantage of reduced construction costs since fewer treatment units are required.

Disadvantages

May be more difficult to operate than conventional activated sludge plant and performance is more dependent on return sludge control.

Aeration tank must be oversized in colder weather conditions resulting in higher costs.

No protection against compounds toxic to nitrifying bacteria.

Submerged Filter

The submerged filter is a biological process in which wastewater flows through a column packed with filter media. Nitrifiers grow on the filter media. The process is best suited to influent ammonia nitrogen levels of 20 mg/l or less. Two stage biological treatment followed by a submerged filter may provide the most economical treatment for a high strength ammonia waste.

Design factors which affect biological nitrification include temperature, dissolved oxygen level, pH, alkalinity, sludge age, and the presence of toxic inhibitors. Nitrifying bacteria work slower in colder winter temperatures and therefore a longer sludge age must be provided. Sufficient dissolved oxygen levels in the range of 1.0 mg/l should be maintained. Optimum pH is 8.0 and should not fall below 7.0. Acidity is produced in the nitrification process, and alkalinity must be supplied to prevent inhibition of nitrification.

Physical-chemical treatment

Selective ion exchange using clinoptilolite, air stripping of ammonia, and breakpoint chlorination are three physical-chemical processes that might be considered for ammonia removal.

Physical-chemical treatment is generally more expensive than biological treatment. It offers the advantage of not being affected by toxic compounds. Low wastewater temperatures may also favor physical-chemical treatment systems due to increased aeration tank volume required by biological treatment. Less land is required and the physical-chemical system is generally more easily controlled than a biological process.

Ammonia stripping

Important operating parameters are pH, air-to-water ratio, and air and sewage temperature. Wastewater pH should be raised to 10.5 to 11.5 before entering the stripping tower. Higher pH and air-to-water ratio should be maintained in colder weather. Advantages include easy operation and the fact that the process may be cost competitive with biological treatment. Disadvantages include scaling problems at cold temperatures, ammonia gas discharged to the atmosphere that may return to the earth, and deterioration of wood packing in tower due to high pH of wastewater.

Ion exchange

Clinopliolite has shown the best selectivity for ammonia removal by ion exchange. Filtration prior to ion exchange is recommended to prevent fouling. pH should be maintained below the range of 7.0 to 8.0 but above 4.0. The system has the disadvantage of lack of a completely satisfactory economic and technical method of recycling the ion exchange regenerant. Pretreatment to remove suspended material is required and organic material in the waste may significantly reduce the resin life.

Breakpoint chlorination

Breakpoint chlorination is the process of adding chlorine to the wastewater until ammonia is oxidized to nitrogen gas. Process variables include chlorine contact time, pH, and the chlorine to ammonia ratio. The optimum chlorine:ammonia nitrogen ratio (by weight) is 7.6:1.0. pH should be controlled at 7.0 or slightly higher for colder temperatures. Breakpoint chlorination offers the advantages of low ammonia nitrogen effluent concentration (less than 1.0 mg/l), an end product of only nitrogen gas, and minimum effect due to change in wastewater temperature. Disadvantages include higher than normal total dissolved solids level in effluent and high chlorine residual.

(b) Ammonia nitrogen removal is desirable to:

1) Reduce the oxygen demand on the receiving stream caused by ammonia nitrogen.

2) Ammonia may be toxic to fish under certain circumstances.

3) Reduce the chlorine demand of the wastewater or any downstream water supplies.

4) Ammonia may react with chlorine to form chloramines that are a less effective disinfectant than free chlorine and may also have toxic effects.

PROBLEM 5.3

A 3.5 mgd wastewater treatment plant treating primarily domestic wastewater is to be designed to achieve a nitrified effluent using a single stage biological treatment process. The plant influent ammonia nitrogen concentration is 18 mg/l and an effluent concentration of 2.0 mg/l is required. The aeration tank of 5-day BOD loading is 5440 lbs/day and the effluent 5-day BOD loading is 800 lbs/day. In order to achieve nitrification during winter operation a solids retention time of 12 days will be used as a design basis. Calculate the required aeration detention time, oxygen requirement, and sludge wasting rate.

Solution

For aeration tank design, calculate the F/M ratio for a solids retention time (SRT) of 12 days using the following equation. Assume Y = 0.65, b = 0.05 $days^{-1}$.

$$\frac{1}{SRT} = Y \frac{F}{M} - b$$

F/M = 0.205 lbs BOD removed/day/lb MLVSS

BOD loading to aeration = 5440 lbs/day

Effluent BOD loading = 800 lbs/day

BOD removed = 4640 lbs/day

F/M = 0.205

$$M = \frac{4640 \text{ lbs}}{0.205} = 22{,}600 \text{ lbs MLVSS}$$

For nitrification assume a MLSS concentration of 2500 mg/l and MLVSS/MLSS ratio = 0.80.

$$\text{Aeration volume} = \frac{22{,}600}{8.34 \times 2500 \times 0.80}$$

$$= 1.35 \text{ million gallons}$$

Calculate detention time at design flow of 3.5 mgd.

$$\frac{1.35 \times 24}{3.5} = 9.3 \text{ hours}$$

Check aeration volumetric loading at design flow.

NH_3-N loading = 3.5 mgd x 8.34 x 18 mg/l

= 525 lbs/day

Aeration volume = 180,500 ft^3

Loading = 2.9 lbs NH_3 - N/1000 ft^3

Oxygen requirement for nitrification will be based on maximum ammonia nitrogen loading. Theoretically, 4.6 pounds of oxygen is required for each pound of ammonia nitrogen removed. For design, a value of 4.3 pounds oxygen/pound ammonia nitrogen removed may be used. A mixed liquor dissolved oxygen level of 2.0 mg/l is also generally recommended for nitrification.

For nitrification,

Assume Maximum flow = 7.0 mgd

NH_3-N removed = 7.0 x (18 - 2) x 8.34 = 934 lbs/day

O_2 required = 934 x 4.3 = 4000 lbs/day

For BOD removal, assume 1.1 lbs/day oxygen required per pound of BOD applied.

5440 x 1.1 = 5985 lbs oxygen for BOD removal

Add 50% safety factor = 9000 lbs/day

Total oxygen required = 4000 + 9000 = 13,000 lbs/day.

Calculate sludge wasting rate for design conditions.

$$\text{Solids retention time} = \frac{\text{Pounds solids under aeration}}{\text{Pounds solids wasted per day}}$$

MLSS under aeration = 1.35 mg x 8.34 x 2500 mg/l = 28,150 lbs.

$$\text{Solids wasted} = \frac{28{,}150 \text{ lbs}}{12 \text{ days}} = 2350 \text{ lbs/day}$$

Assuming effluent solids loading = effluent BOD loading,
Sludge wasted = 2350 - 800 = 1550 lbs/day

Nitrification is temperature dependent. The rate of nitrification decreases below a critical temperature of approximately 15° C, and ceases at a temperature of approximately 5° C. Longer solids retention times are required for design to achieve nitrification requirements for winter operating conditions. It is also important to recognize that alkalinity is destroyed as ammonia is nitrified, indicating the need for pH control, especially for wastewaters with low alkalinity. Approximately 7.14 mg/l of alkalinity as $CaCO_3$ is destroyed per mg/l ammonia nitrogen oxidized.

PROBLEM 5.4

A wastewater treatment plant is to be upgraded to provide nitrification using an attached growth submerged filter as an additional treatment unit. Using the results of laboratory and pilot plant studies design will be based on an ammonia-nitrogen removal rate of 0.008 pounds ammonia nitrogen/day-sq. ft. of filter surface area at a flow rate of 4 gpm/sq.ft. The filter media has a specific surface area of 42 sq. ft./cu. ft. Design flow rate is 500,000 gpd. The untreated ammonia-nitrogen wastewater concentration is 14 mg/l and an effluent concentration of 2 mg/l is required. Assume adequate oxygen is supplied to the process.

(a) For a single pass filter and a filter with a recycle rate of 2:1, calculate the following:

1) Diameter

2) Bed depth

3) Hydraulic detention time

(b) Why might a recycle stream be required?

(c) List the relative advantages and disadvantages of this treatment process.

Solution

(a) Single Pass Filter - consider two filters, each treating 250,000 gpd.

Ammonia removal requirement = 0.25 mgd x 8.34 x (14-2 mg/l) = 25 lbs/day

Calculate filter media surface area and volume required for each filter.

$$\frac{25 \text{ lbs/day}}{8 \times 10^{-3} \text{ lbs/day-ft}^2} = 3{,}125 \text{ ft}^2 \text{ surface area}$$

$$\frac{3{,}125 \text{ ft}^2}{42 \text{ ft}^2/\text{ft}^3} = 74.4 \text{ ft}^3$$

Calculate flow velocity through each filter at a design flow rate of 4 gpm/sq. ft.

$$250{,}000 \text{ gpd} = 174 \text{ gpm} = 0.39 \text{ ft}^3/\text{sec}$$

$$4 \text{ gpm/ft}^2 = 0.00892 \frac{\text{ft}^3/\text{sec}}{\text{ft}^2} = 0.00892 \text{ ft/sec}$$

1) Calculate minimum required filter diameter

$$\text{Area} = \frac{0.39 \text{ ft}^3/\text{sec}}{0.00892 \text{ ft/sec}} = 43.7 \text{ ft}^2$$

Diameter = 7.5 feet

2) Bed depth = $\frac{74.4 \text{ ft}^3}{7.5 \text{ ft}}$ = 10 feet

3) Calculate hydraulic detention time based on volume of filter media required.

$$\frac{74.4 \text{ ft}^3 \times 7.48 \text{ gal/ft}^3}{174 \text{ gal/min}} = 3.2 \text{ minutes}$$

Filter recycle ratio of 2:1 - Consider two filters.

Each filter will treat 750,000 gpd. Assume ammonia nitrogen removal rate remains constant.

Assuming 2 mg/l ammonia nitrogen in recycle stream, ammonia removal requirement is 33 lbs/day.

Filter media surface area = 4,125 ft^2

Filter media volume = 98.2 ft^3

750,000 gpd = 522 gpm = 1.16 ft^3/sec

1) Calculate minimum required filter diameter.

4 gpm/ft^2 = 0.00892 ft/sec

Area = $\frac{1.16 \text{ ft}^3\text{/sec}}{0.00892 \text{ ft/sec}}$ = 130 ft^2

Diameter = 13 feet

2) Bed depth = $\frac{98.2 \text{ ft}^3}{13 \text{ ft}}$ = 8 feet

3) Calculate hydraulic detention time for each filter.

$$\frac{98.2 \text{ ft}^3 \times 7.48}{522 \text{ gpm}} = 1.4 \text{ minutes}$$

(b) A recycle stream is used to reduce concentration of ammonia nitrogen in waste stream and supply oxygen to filter for nitrification.

(c) Advantages

Easy to operate.

May be more cost effective than other treatment processes

if ammonia levels are low.

May be possible to convert to denitrification or add additional units for denitrification without major design changes.

Longer solids retention time at lower hydraulic detention time is possible.

Process adjusts more readily to changes in temperature, pH, and influent ammonia concentrations.

Disadvantages

Filter may clog.

Filter must ordinarily be backwashed.

Does not provide good operation at higher influent waste concentrations.

Possible problem with algae growth on filter packing media.

PROBLEM 5.5

A 0.5 mgd treatment plant for domestic wastewater is to be designed using rotating biological contractors (RBC). The influent BOD_5 concentration is 200 mg/l, influent ammonia nitrogen concentration 20 mg/l, and the winter wastewater temperature may be assumed to be 47° F.

(a) Briefly describe this treatment process.

(b) Calculate the required surface area to meet each of the following effluent requirements:

1) An effluent BOD_5 of 25 mg/l.

2) An effluent ammonia nitrogen concentration of 2.0 mg/l during the summer months.

3) An effluent ammonia nitrogen concentration of 4.0 mg/l during the winter months.

Solution

(a) Rotating biological contactors operate as a fixed-film biological reactor. Circular rotating discs are partially submerged in the waste water to be treated and slowly rotated on a shaft using either a mechanical or air drive system. As the unit rotates a biological growth develops on the media surface for organic removal. Rotation also provides a method of aeration by exposing the unit to the atmosphere. The rotating discs are normally 10-12 feet in diameter and constructed of plastic or other suitable media.

(b) Calculate BOD_5 loading from primary clarifier using <u>Recommended Standards for Sewage Works</u> (Ten State Standards).

Assuming 500 gpd/ft^2 settling rate, BOD removal equals 37%.

Influent BOD concentration to RBC = 200 mg/l x (1-.37)

= 126 mg/l

Effluent BOD required = 25 mg/l

Percent BOD removal required for RBC/final clarifier.

$$\frac{126 - 25}{126} \times 100 = 80\%$$

Percent NH_3 removal required for RBC/final clarifier.

Summer: 90%

Winter: 80%

For the required effluent limitations, BOD influent concentrations, and percent removals calculated above, choose the following design criteria.[3,4]

BOD removal: 5.0 gpd/ft^2 hydraulic loading rate.

Temperature correction factor

@ 47° F = 1.9

NH_3-N removal (Summer): 2.3 gpd/ft^2

NH_3-N removal (Winter): 2.6 gpd/ft^2
Temperature correction factor = 1.6

At temperatures below 55° F both BOD removal rates and nitrification rates decrease and a temperature correction is required. Hydraulic loading rates for temperature below 55° F are adjusted by dividing by the appropriate correction factor.

Design for BOD removal based on winter conditions.

$$\text{Corrected hydraulic loading} = \frac{5 \text{ gpd/ft}^2}{1.9} = 2.63 \text{ gpd/ft}^2$$

$$\text{Total surface area} = \frac{0.5 \times 10^6 \text{ gpd}}{2.63 \text{ gpd/ft}^2} = 190{,}100 \text{ ft}^2$$

Calculate NH_3-N removal requirement for summer and winter conditions.

$$\text{Total surface area: (Summer)} \quad \frac{0.5 \times 10^6 \text{ gpd}}{1.63 \text{ gpd/ft}^2} = 217{,}400 \text{ ft}^2$$

$$\text{Corrected hydraulic loading (winter)} = \frac{2.6 \text{ gpd/ft}^2}{1.6} = 1.63 \text{ gpd/ft}^2$$

$$\text{Total surface area: (Winter)} \quad \frac{0.5 \times 10^6 \text{ gpd}}{1.63 \text{ gpd/ft}^2} = 306{,}800 \text{ ft}^2$$

Considering overall design, a total surface area of 306,800 sq. ft. would be required to achieve nitrification during winter months.

Since the RBC treatment process is dependent on temperature and wastewater characteristics, it is suggested that the equipment manufacturer be consulted regarding design criteria.

PROBLEM 5.6

Biological denitrification to convert nitrate nitrogen to nitrogen gas is achieved under anaerobic conditions by heterotrophic bacteria. Methanol has been commonly used as an organic carbon energy source. Biological denitrification studies using a submerged anaerobic filter to denitrify agricultural tile drainage water indicates that methanol requirements can be determined based on an influent nitrate-nitrogen concentration and "consumptive" ratio as shown in Figure 5.2.[5] Using the results of this study, calculate the methanol requirement for a wastewater with an influent nitrate-nitrogen concentration of 30 mg/l and a winter design temperature of 15° C.

Solution

Using methanol as a carbon source, the consumptive ratio is defined as the ratio of the quantity of methanol required to denitrify and deoxygenate the wastewater plus the methanol required for cell growth divided by the stoichiometric amount of methanol required for denitrification and deoxygenation only. For example, a consumptive ratio of 1.5 indicates 50% excess methanol above the stoichiometric amount is required for cell synthesis. Using a consumptive ratio of 1.3, the following equation was developed by McCarty, et. al., to calculate the actual methanol requirement for denitrification,[6]

$$C_m = 2.47\ N_o + 1.53\ N_1 + 0.87\ D_o \qquad \text{(Eqn. 1)}$$

where C_m = required methanol concentration, mg/l

N_o = influent NO_3-N concentration, mg/l

N_1 = influent NO_2-N concentration, mg/l

D_o = dissolved oxygen concentration of the influent wastewater stream, mg/l

For a consumptive ratio of 1.0, equation (1) may be written as,

$$C_m = 1.9\ N_o + 1.18\ N_1 + 0.67\ D_o \qquad \text{(Eqn. 2)}$$

From Figure 5.2, for an influent nitrate-nitrogen concentration of 30 mg/l and a temperature in the range of 12° C to 16° C, the consumptive ratio would be approximately 1.7. (This ratio would drop to 1.1 for a temperature of 20° C to 24° C.) Assume a nitrite-nitrogen concentration of 0.5 mg/l and a dissolved oxygen concentration of 4.0 mg/l. Equation (2) may be written as,

$$C_m = 3.23\ N_o + 2.0\ N_1 + 1.1\ D_o$$

The methanol requirement is calculated as,

$$C_m = 3.23\ (30\ \text{mg/l}) + 2.0\ (0.5\ \text{mg/l}) + 1.1\ (4.0\ \text{mg/l})$$

$$C_m = 102.3\ \text{mg/l}$$

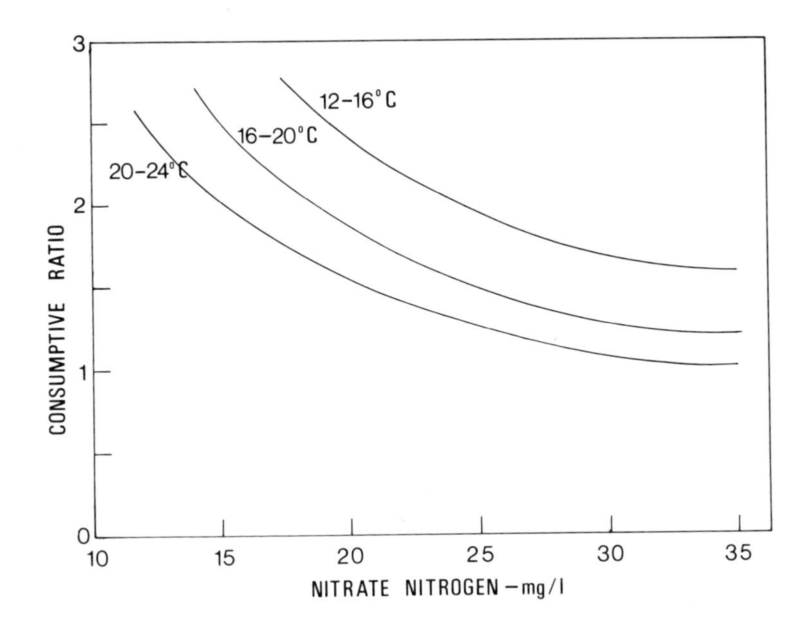

Figure 5.2. Comsumptive ratio as a function of temperature and nitrate nitrogen concentration.

The ratio of methanol/nitrate-nitrogen or M/N ratio can be calculated as,

$$\frac{102.3\ mg/l}{30\ mg/l} = 3.41$$

Therefore, 3.41 pounds methanol would be required per pound of nitrate-nitrogen removed. Figure 5.3 shows a graph of M/N ratio versus influent nitrate-nitrogen concentration as a function of temperature. Using this graph, methanol requirements for biological denitrification can be determined.

PROBLEM 5.7

Carbon adsorption isotherm tests are run on carbon A and carbon B using laboratory beakers. Pulverized carbon is added in different amounts to seven 500 ml samples of wastewater, the samples are agitated for two hours, and the soluble Total Organic Carbon (TOC) concentration is measured. Results are shown in Table 5.3. Determine which carbon you would recommend for further carbon column studies based on the results of this screening test.

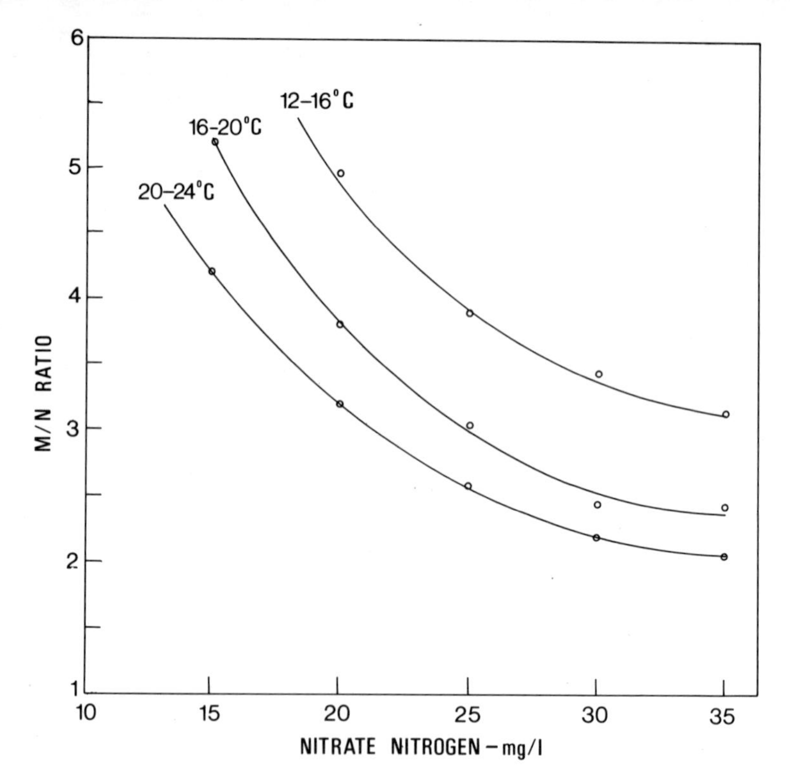

Figure 5.3. Methanol requirements for denitrification.

Solution

Using a carbon adsorption isotherm, the amount of TOC adsorbed at equilibrium and the rate at which equilibrium is attained can be determined and the appropriate carbon selected.

The Freundlich equation is most often used to describe the adsorption isotherm,

$$\frac{X}{M} = kC^{1/n}$$

where X = amount of organic adsorbed

M = weight of carbon

C = equilibrium concentration of unadsorbed organic in solution

k, n = constants

Table 5.3

TOC Equilibrium Data

Weight Carbon Added (grams)	Carbon Dose (gm/l)	TOC Concentration at Equilibrium (mg/l) Carbon A	Carbon B
0	0	144	144
0.05	0.1	135	132
0.10	0.2	128	119
0.50	1.0	95	55
1.0	2.0	58	27
2.0	4.0	30	12
5.0	10.0	12	5

Table 5.4

Carbon Adsorption Isotherm Data

Carbon A		Carbon B	
TOC (mg/l)	mg TOC / gm Carbon	TOC (mg/l)	mg TOC / gm Carbon
135	90	132	120
128	80	119	125
95	49	55	89
58	43	27	59
30	29	12	33
12	13	5	14

Taking the log of both sides of this equation yields a linear equation with slope = 1/n and intercept k at C = 1 mg/l,

$$\log \frac{X}{M} = \log k + \frac{1}{n} \log C$$

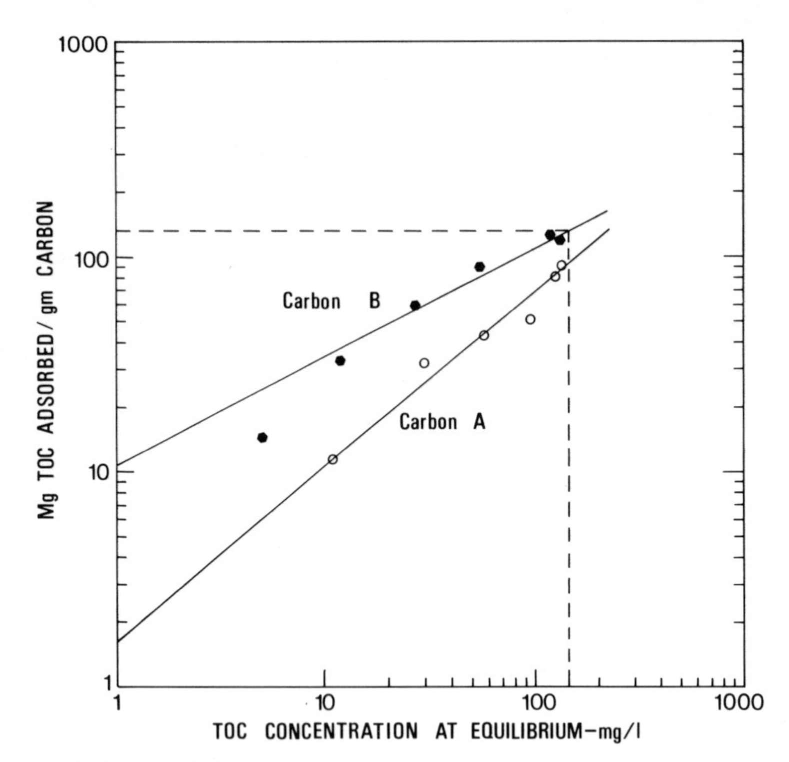

Figure 5.4. Carbon adsorption isotherm for carbon A and carbon B.

The amount of TOC adsorbed per gram of carbon is calculated for each residual TOC concentration. Results are summarized in Table 5.4 and shown in Figure 5.4. By extrapolating each curve to the initial TOC concentration of 144 mg/l, the removal capacity of carbon A is estimated as 93 mg TOC/gm carbon and carbon B as 130 mg TOC/gm carbon. Carbon B is selected. Values of k and n are determined from Figure 5.4 as follows.

Carbon A: $k = 1.6$
$1/n = 0.81$
$n = 1.24$

Carbon B: $k = 11.0$
$1/n = 0.56$
$n = 1.80$

In most cases several different carbons are evaluated and one or more are selected for continuous-flow column testing or further isotherm tests. Figure 5.5 show TOC adsorption isotherms for eight different carbons. The dashed line

Table 5.5

Carbon Efficiency

Carbon	$\frac{X}{M}$	Relative Efficiency %
1	400	100
2	270	68
3	150	38
4	100	25
5	60	15
6	6	2
7	20	5
8	9	2

represents the average influent waste concentration for the tests that were made. Each isotherm is extrapolated to this line and the X/M value at that point is determined. The relative adsorption efficiency is determined by making the largest X/M value equal to 100% and calculating other carbon efficiencies as a percentage of the largest value. Results are shown in Table 5.5 with carbon number 1 extrapolated to an X/M value of 400 and taken as 100%.

Adsorption isotherms should be evaluated relative to influent waste concentration, required effluent concentration, and carbon column mode of operation. For example, if it is assumed that each carbon is required to produce an effluent concentration of 20 mg/l, carbon 6 would not meet this requirement. Carbon 8 is non-linear and does not fit the Freundlich isotherm equation, indicating the likely presence of non-adsorbable compounds. Carbons 1 and 2 exhibited the highest removal efficiencies and reduced the influent concentration to less than 10 mg/l, while carbon 3 produced the lowest residual concentration over a wider concentration range. Carbons 1, 2, and 3 would likely be considered for further evaluation. In a fixed bed column operation a carbon with an adsorption capacity over a wide concentration range range becomes important since the carbon will be removed infrequently. In a countercurrent moving bed column design, a carbon with a high removal capability at the influent waste concentration would generally be more effective.

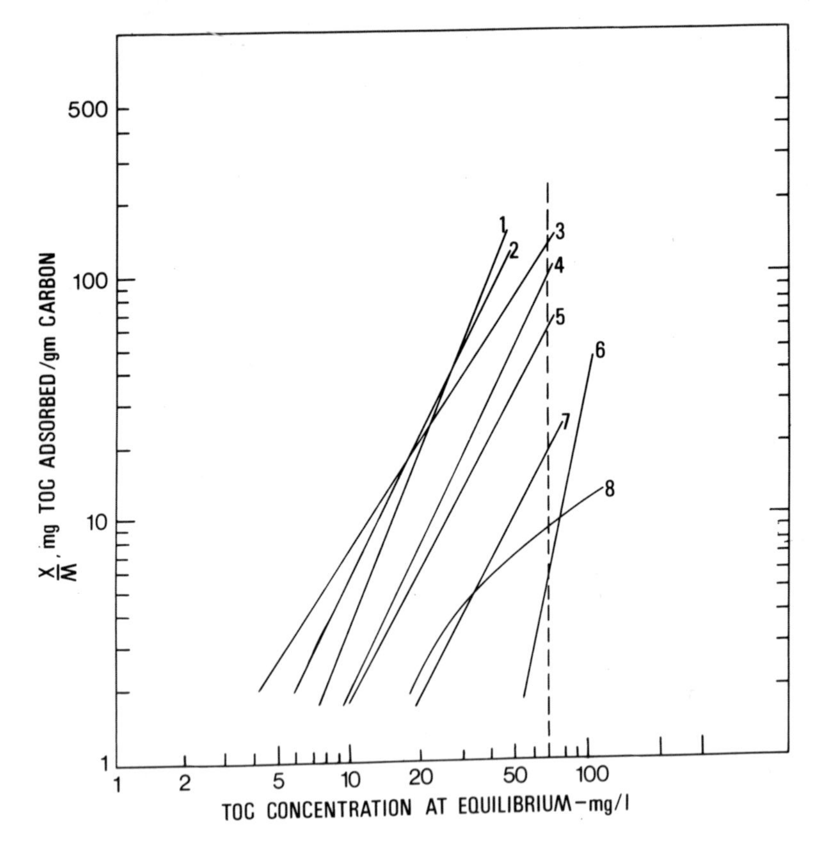

Figure 5.5. Carbon adsorption isotherms.

PROBLEM 5.8

Using the carbon isotherm in Problem 5.7, calculate the theoretical amount of carbon B required to reduce an initial TOC concentration of 100 mg/l to 10 mg/l in a single stage and a two stage carbon contactor.

Solution

The equation for the Freundlich isotherm may be written as,

$$\frac{C_o - C_f}{M} = kC_f^{1/n}$$

where M = carbon weight, gm/l

C_o = initial TOC concentration, mg/l

C_f = residual TOC concentration, mg/l

k = intercept at $C_f = 1$, mg/gm

$1/n$ = slope

For a single stage carbon contactor,

$$\frac{100 - 10}{M} = (11)(10)^{0.56}$$

$$M = 2.25 \text{ gm/l}$$

$$M = 2250 \text{ mg/l}$$

For a two stage carbon contactor, assume 50% TOC removal in the first column or an influent TOC concentration of 50 mg/l to the second column.

For the first stage column,

$$\frac{100 - 50}{M} = (11)(50)^{0.56}$$

$$M = 508 \text{ mg/l}$$

For the second stage column,

$$\frac{50 - 10}{M} = (11)(10)^{0.56}$$

$$M = 1000 \text{ mg/l}$$

Total carbon = 1508 mg/l

By using more than one stage of treatment, carbon requirements are reduced from 2250 mg/l to 1508 mg/l.

PROBLEM 5.9

In a complex industrial manufacturing plant a particular waste stream from a manufacturing area is to be treated using activated carbon adsorption. The process is expected to be in operation a maximum of 12 days each month. At the end of the operating period the carbon will be removed and replaced with fresh carbon or regenerated, depending on a further economic analysis. The following design data will be used:

Carbon contactor	Single stage fixed bed
Flow rate	275 gpm
Hydraulic loading	2.5 gpm/sq. ft.
Effluent COD	1200 mg/1
Influent COD	2252 mg/1

Figure 5.6 shows the breakthrough curves from a carbon column pilot study treating a waste considered representative of actual plant operation. Four beds of activated carbon were used in series, each bed six feet in depth and six inches in diameter. Each bed contained 12 x 40 mesh activated carbon. Flow rate was 97 liters per hour. Calculate the carbon bed depth, carbon usage rate, and theoretical carbon capacity.

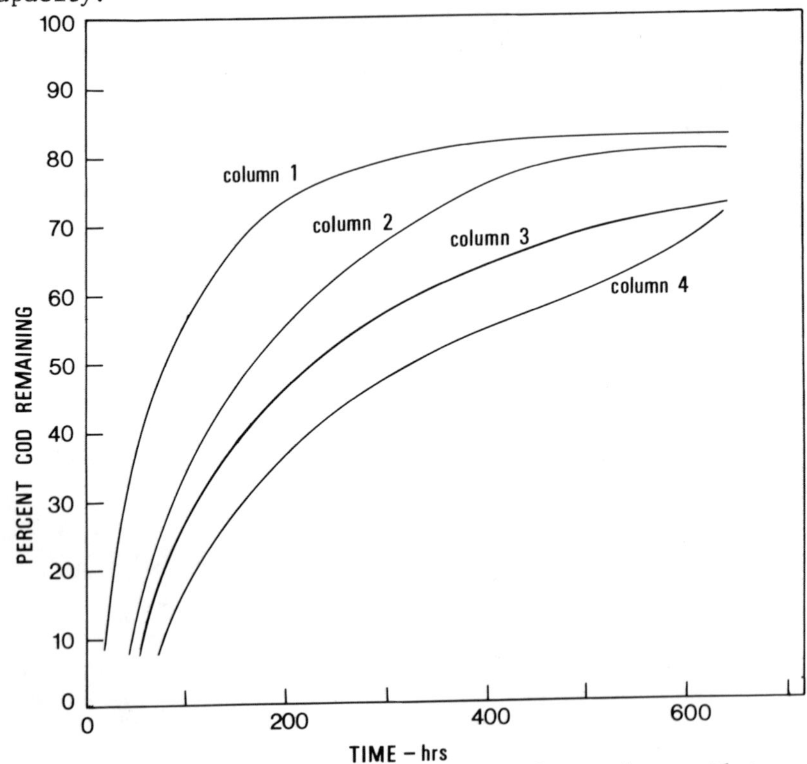

Figure 5.6. Breakthrough curves for carbon column pilot study. Problem 5.9.

Solution

Problem will be solved using the Bohart-Adams equation in the form of Bed Depth Service Time (BDST) as described by Hutchins.[7] The Bohart-Adams equation may be written as,

$$t = \frac{N_o}{C_o V} X - \frac{1}{KC_o} \ln [(C_o/C_B) - 1]$$

where t = service time at breakthrough, hours

X = bed depth, feet

V = hydraulic loading, or linear velocity, ft/hr

C_o = influent waste concentration, lb/ft^3

C_B = effluent waste concentration, lb/ft^3

N_o = adsorption efficiency, lb/ft^3 carbon

K = adsorption rate constant, $ft^3/lb\text{-}hr$

For the breakthrough curve in Figure 5.6, an effluent concentration of 1200 mg/l is equivalent to 53% COD remaining, or 47% COD removal. Breakthrough points for each column at 47% COD removal are as follows:

Column 1	90 hours
Column 2	190 hours
Column 3	250 hours
Column 4	375 hours

A graph is made of service time versus bed depth as shown in Figure 5.7 using these four data points. This straight line is in the form of the Bohart-Adams equation with,

$$\text{slope} = \frac{N_o}{C_o V} = 15.8 \text{ hr/ft}$$

$$\text{intercept} = \frac{1}{KC_o} \ln [(C_o/C_B) - 1] = - 10 \text{ hours}$$

Hutchins suggests that when using the BDST equation, the flowrate used in pilot tests may be applied to design without significant error by multiplying the slope (15.8 hr/ft) by the ratio of the two flow rates.

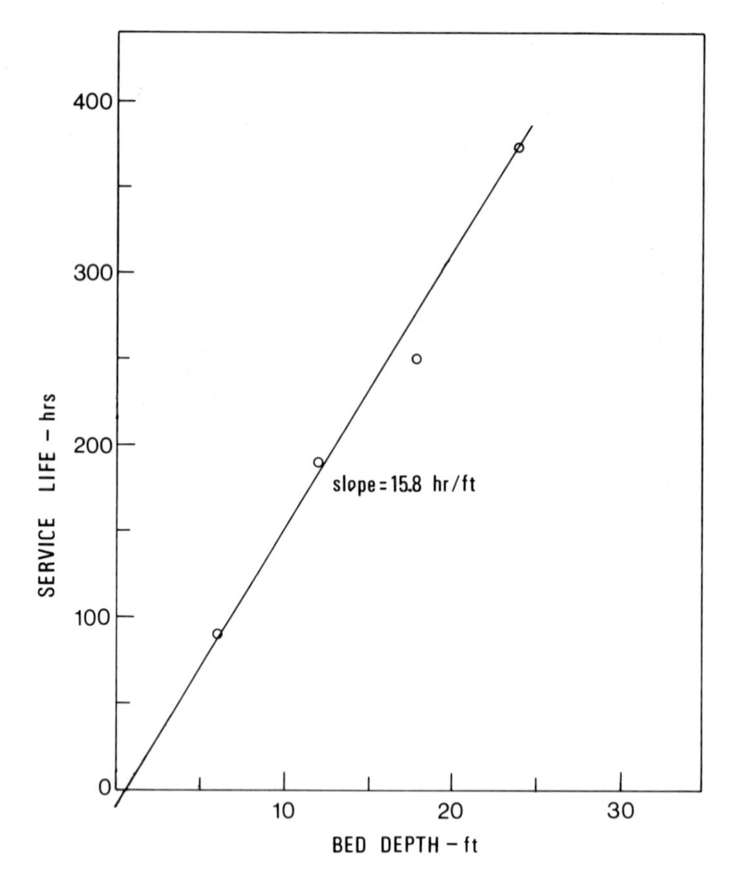

Figure 5.7. Service time versus bed depth.

From pilot study, 97 liters/hour = 2.2 gpm/ft^2

$$15.8 \text{ hr/ft} \times \frac{2.2 \text{ gpm/ft}^2}{2.5 \text{ gpm/ft}^2} = 13.9 \text{ hr/ft}$$

BDST = 288 hours

$$t = 13.9 (X) - 10$$

$$X = \frac{288 \text{ hrs} + 10 \text{ hrs}}{13.9 \text{ hr/ft}} = 20.8 \text{ feet bed depth}$$

Calculate carbon required. Use carbon density of 25 lb/cu. ft.

$$\text{Column bed area} = \frac{275 \text{ gpm}}{2.5 \text{ gpm/ft}^2} = 110 \text{ ft}^2$$

Diameter = 11.8 feet

$110 \text{ ft}^2 \times 20.8 \text{ ft} \times 25 \text{ lb/ft}^3 = 57{,}200$ lbs carbon

$$\text{Carbon usage rate} = \frac{57{,}200 \text{ lb}}{288 \text{ hrs}} = 199 \text{ lb/hr}$$

Calculate carbon capacity based on both pounds COD applied and pounds COD removed.

Flow rate = 275 gpm = 0.4 mgd

COD applied = 2250 mg/l x 8.34 x 0.4 mgd

= 7,500 lbs/day

COD removed = (2250-1200) x 8.34 x 0.4

= 3500 lbs/day

$$\text{Carbon capacity (COD removed)} = \frac{3{,}500 \text{ lbs COD}}{57{,}200 \text{ lbs carbon}}$$

= 0.06 lbs COD removed/lb carbon

$$\text{Carbon capacity (COD applied)} = \frac{7{,}500 \text{ lbs COD}}{57{,}200 \text{ lbs carbon}}$$

= 0.13 lbs COD applied/lb carbon

It should be noted that the BDST equation as used in this problem is applicable only to a single stage fixed bed column. However, Hutchins suggests methods to modify the equation using correction factors to design for a series of columns or a moving bed operation.

In conducting column studies, it is recommended that carbon which has been regenerated a number of times be used since the carbon adsorption properties may decrease significantly after regeneration, resulting in a system that is underdesigned with increased operating costs.

PROBLEM 5.10

Activated carbon adsorption is being considered for treatment of an industrial waste. Pilot plant treatability studies are conducted using 3 carbon columns in series. Each column is 10 inches in diameter and 15 feet deep. Approximately 195 pounds of regenerated activated carbon is added to each column. At a feed rate of 3 gallons/minute, column breakthrough occurs after 32 days of operation. Using this data, develop the design criteria for a plant treating 2.0 mgd of wastewater. Assume that three counter-current beds in series will be used with one stand-by contactor. When breakthrough occurs, the first contactor will be taken offstream and the carbon regenerated. Carbon bed depth in each contactor will be 15 feet and the same hydraulic loading rate used in the pilot studies will be used for design.

Solution

Evaluate Pilot Plant Operation

For 3 columns in series, each column 10 inches x 15 feet deep,

Column area = 0.545 ft^2

$$\text{Hydraulic loading} = \frac{3\ \text{gpm}}{0.545\ \text{ft}^2} = 5.5\ \text{gpm/ft}^2\text{/column}$$

Use 5 gpm/ft^2 for design.

$$\text{Carbon dosage} = \frac{3\ \text{columns x } 195\ \text{lbs/column}}{3\ \text{gpm x } 1440 \text{ x } 32\ \text{days}}$$

$$= 4.2\ \text{lbs carbon/1000 gallons treated}$$

$$= 238\ \text{gallons/lb carbon}$$

$$\text{Contact time (per column)} = \frac{\text{bed volume}}{\text{flow rate}}$$

$$= \frac{0.545\ \text{ft}^2 \text{ x } 15\ \text{ft}}{3\ \text{gpm}/7.48}$$

Contact time = 20.4 minutes/column

Develop Design Criteria

Design flow rate = 2.0 mgd

Loading rate = 5 gpm/ft^2

Three columns in series are recommended as in pilot study. Use 10 foot diameter columns with 15 foot bed depth. Allowing 50% bed expansion, minimum contactor vessel depth is 23 feet. Use approximately same contact time as in pilot study.

Bed area = 78.5 ft^2

Bed volume = 1177 ft^3

$5\ gpm/ft^2 \times 78.5\ ft^2 = 392.5\ gpm = 565{,}200\ gpd$

Flow rate of 565,200 gpd through 3 columns in series will require 4 parallel trains for a design flow rate of 2.0 mgd.

Total contactors (including standby) = 16

Calculate contact time at 2.0 mgd or 347 gpm through 3 columns in series.

$$\text{Contact time} = \frac{1177\ ft^3 \times 7.48\ gal/ft^3}{347\ gpm}$$

$$= 25.4\ \text{minutes/column}$$

Calculate total pounds of carbon required based on 12 columns in operation. Assume a carbon density of 25 lbs/cu. ft.

$12 \times 1177\ ft^3 \times 25\ lb/ft^3 = 353{,}100\ lbs$

Calculate bed life until column breakthrough using carbon dosage rate from pilot plant study and compare with pilot plant breakthrough time.

$$\frac{353{,}100\ \text{lbs carbon} \times 238\ gal/lb}{2.0 \times 10^6\ gal/day} = 42\ \text{days}$$

Pilot plant breakthrough time is 32 days. Select a bed replacement interval of 35 days for design.

Calculate carbon exhaustion rate for each train based on 35 day replacement interval.

$$\frac{353{,}100/4}{35} = 2522\ lb/day = 105\ lb/hr$$

Determine regeneration furnace loading. Assume 30 percent downtime.

$$\frac{105\ lb/hr}{0.7} = 150\ lb/hr$$

PROBLEM 5.11

Identify two treatment methods for upgrading a primary treatment plant using a physical-chemical treatment process with powdered activated carbon addition, and two treatment methods using a combined biological and powdered activated carbon in conjunction with physical-chemical or biological wastewater treatment.

Solution

Two possible treatment schemes for upgrading a primary treatment plant using chemical addition and powdered activated carbon are shown in Figure 5.8 (a) and 5.8 (b). In Figure 5.8 (a) the influent wastewater is treated in a single stage powdered activated carbon contacting unit, followed by chemical addition of a metal salt, such as alum, ferric chloride, or lime, flocculation and clarification. The effluent from the clarifier may be filtered using a granular media filter, depending upon the desired effluent quality. Settled solids from the clarifier are recycled to the carbon contractor or are wasted. A possible variation would be to use chemical treatment ahead of the carbon contactor. Tube settlers may be added to the clarifier to improve solids capture.

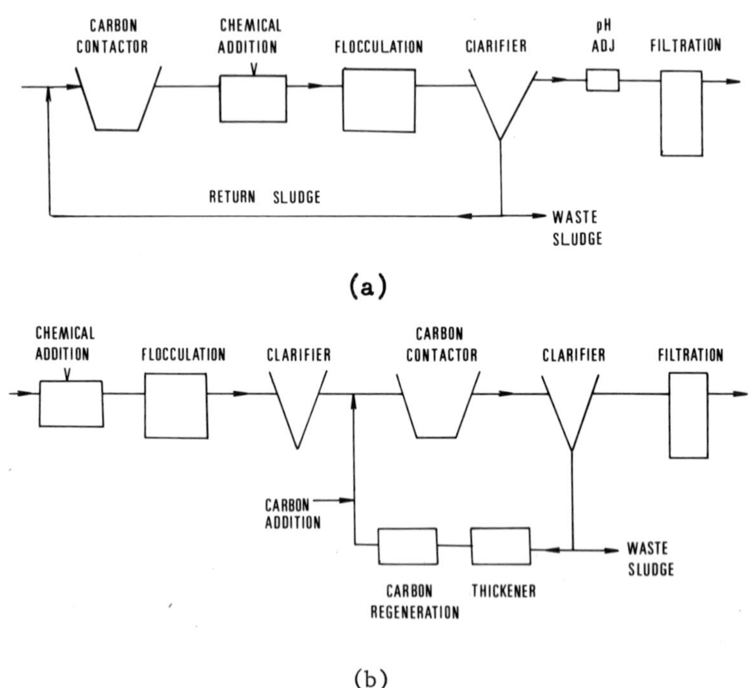

Figure 5.8 (a) Flow scheme for powdered carbon addition preceding chemical treatment. (b) Flow scheme for powdered carbon addition with separate stage carbon contactor, clarifier, and regeneration.

Figure 5.8 (b) also shows a single-stage powdered carbon treatment process in which a separate carbon contactor and clarifier follows a chemical treatment process. This flow scheme offers the advantage of regeneration of the spent carbon. Filtration would be recommended for treatment of the effluent prior to discharge.

Figure 5.9 (a) and 5.9 (b) show two possible flow schemes for powdered activated carbon addition to an activated sludge process. Neither process requires additional treatment units except for carbon regeneration and sludge handling. In Figure 5.9 (a) carbon is added to the aeration tank or alternately to the final clarifier. Sludge from the final clarifier may be thickened and the carbon regenerated.

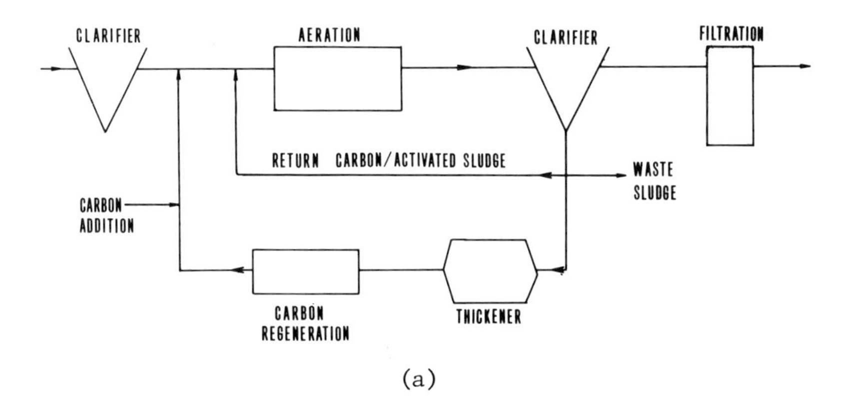

(a)

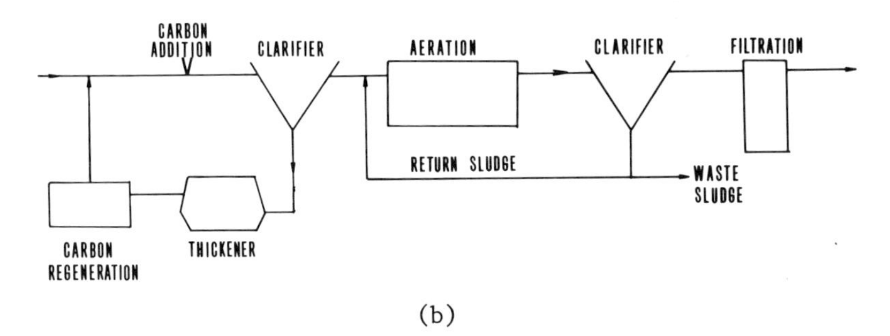

(b)

Figure 5.9. Flow scheme for powdered carbon addition to biological treatment process. (a) Aeration unit (b) Primary clarifier.

In Figure 5.9 (b) powdered carbon is added to the primary clarifier ahead of the biological treatment process. This offers the possible advantage of removing any toxic pollutants that may be inhibitory to the biological process. The primary sludge may also be regenerated. For either treatment scheme, effluent filtration may also be desirable.

The following advantages have been reported for powdered activated carbon addition:[8]

(a) Removal of color and odor.

(b) Removal of adsorbable organic compounds and increased stability against toxic pollutant loadings.

(c) Improved sludge settling and dewatering characteristics.

(d) Improved operation and effluent quality.

PROBLEM 5.12

Thirty samples of a wastewater treatment plant effluent are analyzed for both 5-day BOD and chemical oxygen demand (COD). The data is shown in Table 5.6. Determine if there is a statistical correlation between BOD and COD based on this data.

Solution

Calculate the correlation coefficient, r. The correlation coefficient is a measure of the relationship between two variables. For this example, the closer the value of r is to +1, the greater the degree of correlation. To determine if COD can be used to estimate BOD, COD is defined as the independent "x" variable and BOD the dependent "y" variable. The correlation coefficient is a useful statistical test when neither the x nor y value is a controlled sampling parameter. In contrast, a regression analysis assumes that the independent value is chosen with a preassigned value.

Using a desk calculator, the correlation coefficient r is determined to be 0.9675. Using statistical tables to test for independence, it may be concluded there is a linear relationship between BOD and COD at better than a 99.9% confidence level.

A linear regression analysis of the data results in the equation,

BOD = 0.4707 COD - 13.04

The regression line and data is shown in Figure 5.10.

Table 5.6

Data for Problem 5.12

COD (lbs/day)	BOD_5 (lbs/day)
494	216
444	200
528	238
396	164
532	230
308	116
350	150
456	190
440	190
544	248
310	120
538	226
480	200
500	222
396	176
486	202
556	240
600	280
428	184
440	194
291	134
490	215
546	246
582	292
368	177
386	193
400	165
347	160
278	125
304	137

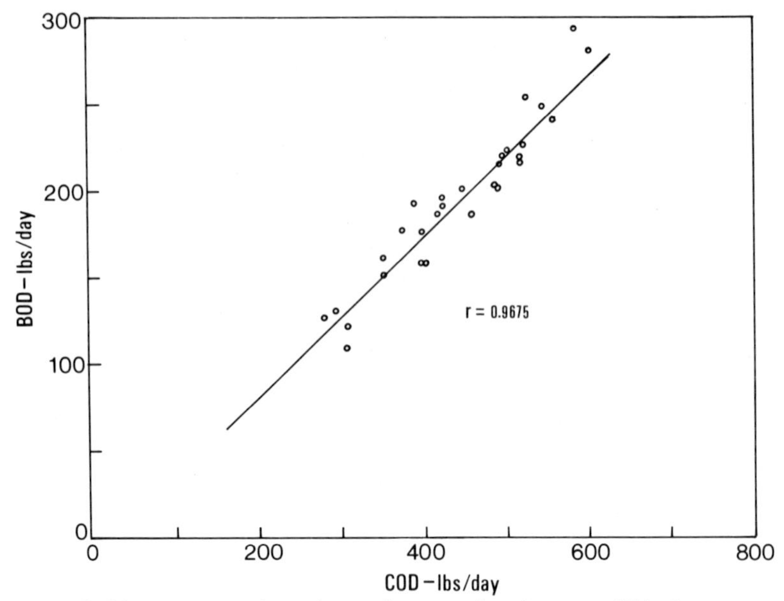

Figure 5.10. Determination of correlation coefficient, r, for Problem 5.12.

PROBLEM 5.13

For the data given in Problem 5.12, determine the BOD and COD loading that will be exceeded 50% and 90% of the time.

Solution

The data given in Problem 5.12 is first organized in increasing order of magnitude as shown in Table 5.7. This organizes the data so that it can be assigned a "plotting position" obtained from statistical tables for a sample size of 30. The data is then plotted on normal (rectangular coordinate) probability paper to determine the statistical variation as shown in Figure 5.11.

Figure 5.11 shows that BOD will exceed 195 lbs/day 50% of the time, and COD will exceed 440 lbs/day 50% of the time. A BOD value of 260 lbs/day can be achieved 90% of the time, and a COD value of 580 lbs/day can be achieved 90% of the time. Conversely stated, a BOD of 260 lbs/day and COD of 580 lbs/day will be exceeded only 10% of the time. The 50% probability point is the mean or average value of the data.

Probability plots are extremely useful to organize data and determine fluctuations and variations in hydraulic loading, influent and effluent data, and other plant operating parameters. Data may be plotted on log-normal as well as normal probability paper.

Table 5.7

Data for Probability Plot

BOD (lbs/day)	COD (lbs/day)	Plotting Position
116	278	2.1
120	291	5.3
125	304	8.7
134	308	11.9
137	310	15.2
160	347	18.7
164	350	21.8
165	368	25.1
176	386	28.4
177	396	31.9
184	396	35.2
190	400	38.6
190	428	41.7
193	440	45.2
194	440	48.4
200	444	51.6
200	456	54.8
202	480	58.3
215	486	61.4
216	490	64.8
222	494	68.1
226	500	71.6
230	528	74.9
238	532	78.2
238	538	81.3
240	544	84.8
246	546	88.1
248	556	91.3
280	582	94.7
292	600	97.9

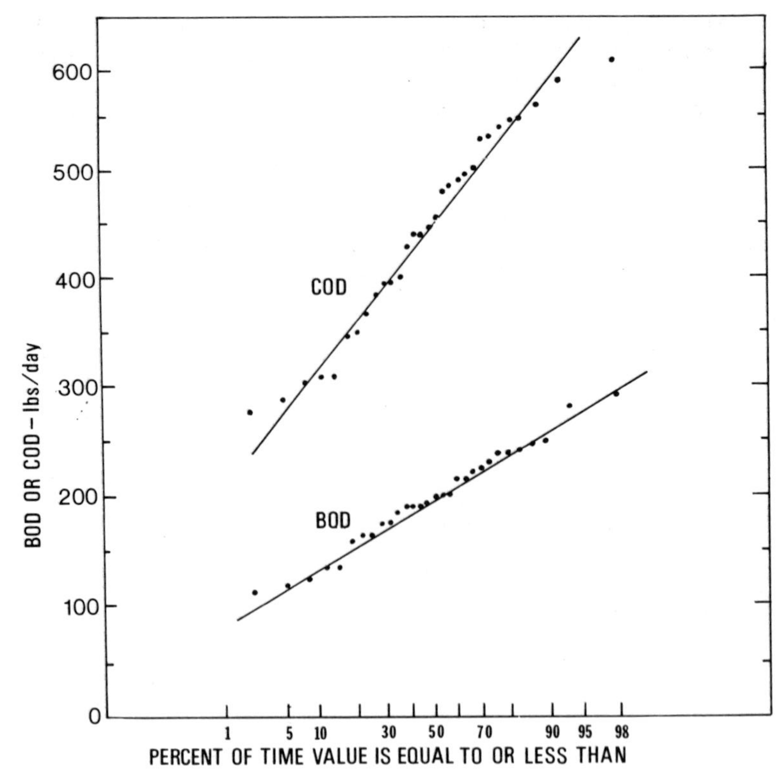

Figure 5.11. Probability plot of data from Problem 5.12.

PROBLEM 5.14

In Table 5.8 sludge wasting rate (lbs/hr) and mixed liquor suspended solids (MLSS) level is tabulated in columns 2 and 4 during a 32 day operating period for an activated sludge plant. Determine if there is a trend between changes in sludge wasting rate and MLSS concentration.

Solution

This problem is evaluated by determining the 7-day moving average for both the sludge wasting rate and MLSS concentration. The 7-day moving average for any given day is the average of the data for that day plus the six previous days. This is tabulated in columns 3 and 5 of Table 5.8. A 7-day moving average can be used to determine the effect that changing one operating variable has on plant performance or another operating variable.

For example, Figure 5.12 shows the data for sludge wasting rate and MLSS plotted without using a 7-day average. Because of the large daily variations, the data is difficult to evaluate. The data is plotted again in Figure 5.13, this time using a 7-day moving average to smooth out variations in the data. A general trend of decreasing MLSS values with decreasing sludge wasting rate is shown.

This type of analysis is useful since many operating changes do not affect plant operation until several days after the change is made. The use of 28-day moving averages for long term data evaluation may also be used.

Table 5.8

Data for Problem 5.14

Day	Sludge Wasted (lb/hr)	7-Day Average	MLSS (mg/l)	7-Day Average
1	1167		4294	
2	735		5404	
3	554		3784	
4	1283		3659	
5	840		3642	
6	998		3569	
7	991	938	3645	4000
8	997	914	3829	3933
9	840	929	4029	3736
10	1050	1000	3493	3695
11	1219	990	3272	3640
12	702	971	3175	3572
13	500	890	3341	3541
14	500	830	3570	3530
15	750	794	3867	3536
16	735	779	3387	3444
17	630	719	3603	3459
18	443	608	3349	3470
19	771	618	3100	3460
20	420	607	3162	3434
21	887	662	3922	3485
22	630	645	3587	3445
23	662	635	3233	3423
24	816	661	2963	3331
25	838	718	3284	3322
26	735	713	2904	3294
27	221	684	3361	3321
28	1050	707	3464	3257
29	1050	767	3599	3258
30	420	733	3546	3303
31	840	736	3669	3404
32	776	727	3553	3442

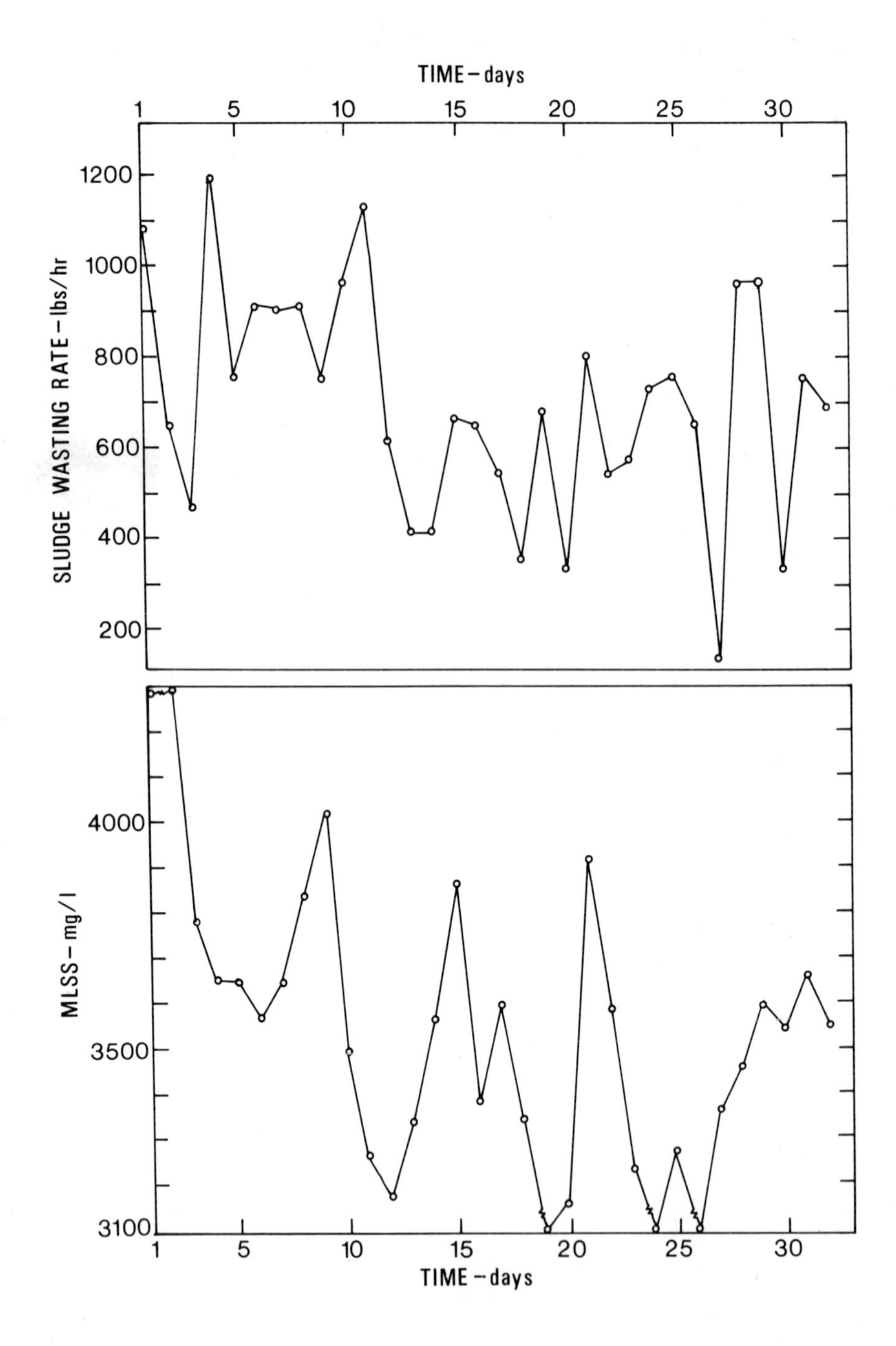

Figure 5.12. Sludge wasting rate and MLSS versus time for data in Problem 5.14.

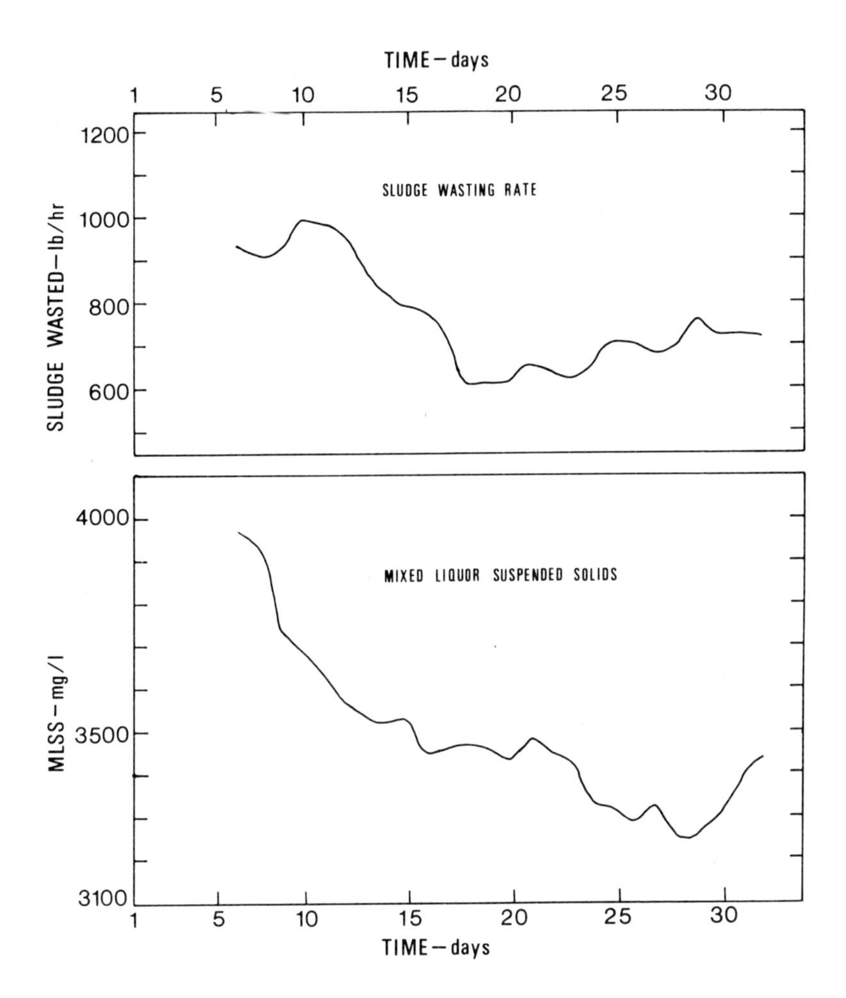

Figure 5.13. Graph of 7-day moving average for data shown in Figure 5.12.

PROBLEM 5.15

Results of a 96-hour continuous flow bioassay test are summarized in Table 5.9. Determine the 96-hour tolerance limit (TL_{50}) based on this data.

Solution

Bioassays are used to evaluate wastewater toxicity using

Table 5.9

Bioassay Test Results

% Wastewater Volume	Number of Test Organisms	Number of Live Organisms
Control	20	20
4.5	20	19
8.0	20	18
15.0	20	13
25.0	20	11
45.0	20	6
70.0	20	1
100.0	20	0

fish as the test organism. The TL_{50} is equivalent to the median tolerance limit and is defined as that concentration of a substance in which 50% of the fish survive for 96 hours. Bioassay results are commonly reported as LC_{50} (lethal concentration) which is analagous to TL_{50}.

Figure 5.14 shows a graph of percent wastewater by volume versus the percent survival. A straight line is drawn between the data points directly above and below the 50% survival point. The 96-hour TL_{50} is estimated as 27% from the graph.

PROBLEM 5.16

Table 5.10 shows the results of a granular media filtration pilot plant study using effluent from a trickling filter plant. Each filter run was terminated at a head loss of 10 feet. A statistical analysis of the trickling filter plant effluent data indicates that an effluent suspended solids concentration of 40 mg/l is achieved 90% of the time. A filter effluent of 20 mg/l suspended solids is desired. Calculate the surface area required for filtration. Average daily plant flow rate is 3.0 mgd, maximum daily flow rate is 5.8 mgd, and maximum flow rate over a 4-hour period is 7.2 mgd based on plant operating records. Assume 5% of the average daily flow is required for backwashing each filter.

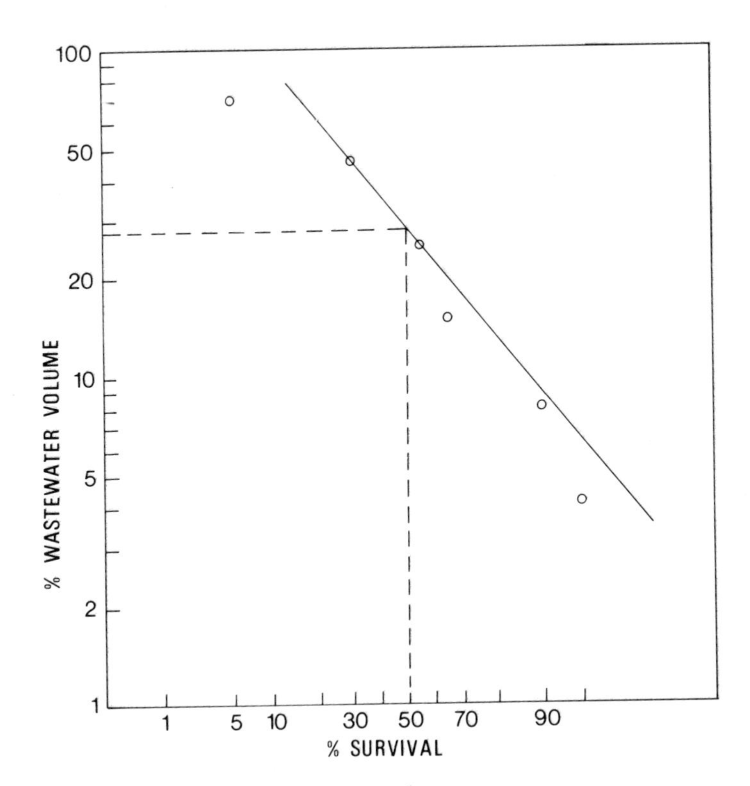

Figure 5.14. Bioassay test results.

Solution

Because the effluent quality and flow rate can be expected to vary considerably from the treatment plant, the data from Table 5.10 is divided into three categories of influent suspended solids concentration to the pilot plant filter. Filter removal efficiency and length of filter run is shown as a function of hydraulic loading in Figure 5.15. For a suspended solids concentration of 40 mg/l (30 mg/l to 50 mg/l range) an effluent suspended solids concentration of 20 mg/l (50% to 55% removal) can be expected at a flow rate of 3.0 to 4.0 gpm/sq. ft. The filter run time would be in the range of 25 to 33 hours. For design, a flow rate of 3.5 gpm/sq. ft. will be used and it will be assumed that filters will be backwashed every 24 hours. Filtration area is calculated based on maximum 4-hour flow rate plus amount of backwash water required. Backwash water will be recirculated back to the treatment plant.

Table 5.10

Wastewater Filtration Pilot Plant Data

Influent Suspended Solids (mg/1)	Effluent Suspended Solids (mg/1)	Filter Loading (gpm/ft^2)	Bed Life (hours)
65	29	2.0	33
37	14	2.5	39
23	13	4.5	20
46	21	2.8	43
51	29	2.8	23
23	10	2.8	38
22	13	3.6	28
16	10	4.3	36
28	13	4.5	28
39	20	4.8	20
102	77	5.1	14
82	45	5.2	16
46	23	5.5	16
51	31	5.9	16
54	27	6.2	20
46	17	1.5	49
49	25	2.6	30

Backwash water = 3.0 mgd x 0.05 = 0.15 mgd

$$\text{Area} = \frac{7.35 \times 10^6 \text{ gal/1440 min}}{3.5 \text{ gpm/ft}^2} = 1460 \text{ ft}^2$$

Add 20% safety factor to account for period when filter is out of service for backwashing or maintenance.

Filtration area = 1750 ft^2

Use four filters, each 440 ft^2

Check filtration rate at maximum flow with one filter out of service.

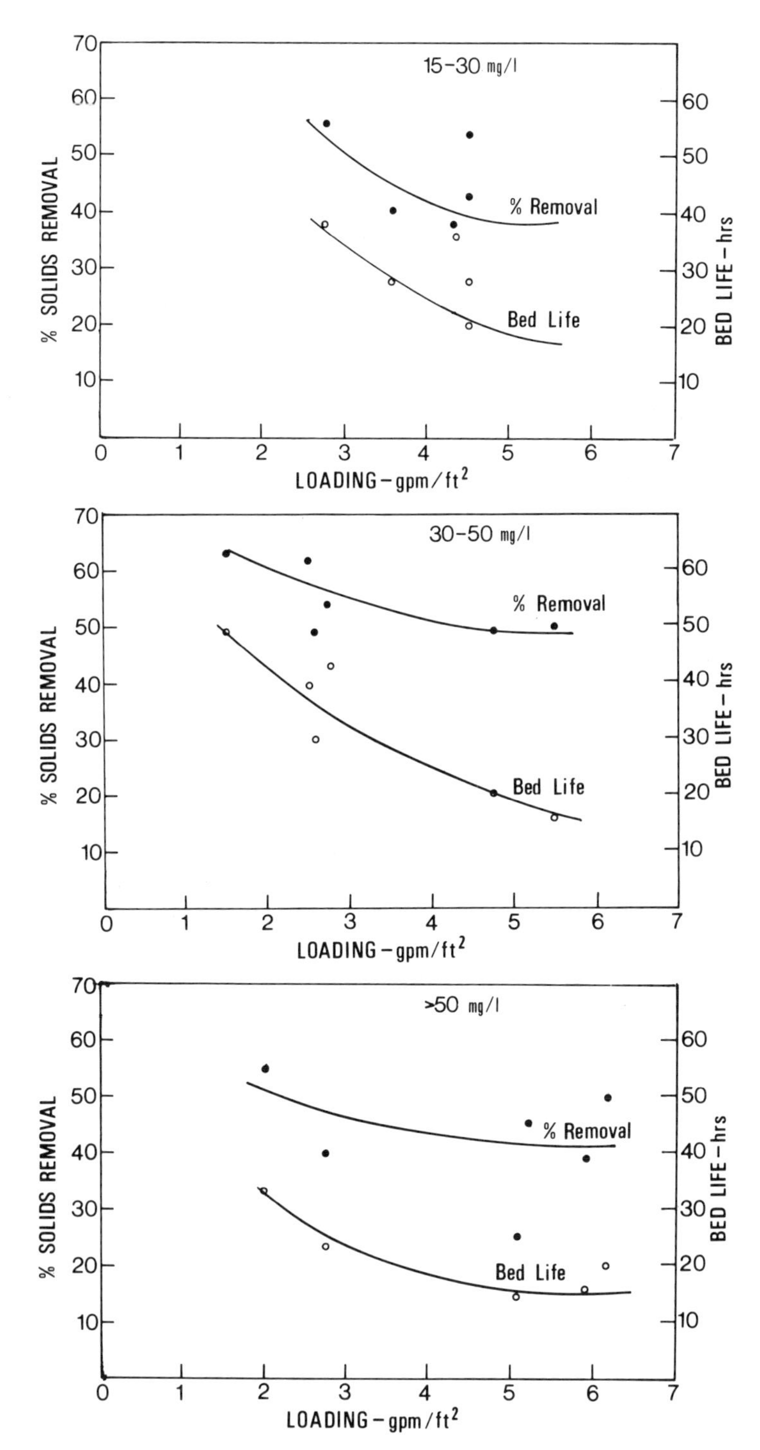

Figure 5.15. Granular media filtration data.

$$\frac{7.2 \times 10^6 \text{ gal/1440 min}}{3 \times 440 \text{ ft}^2} = 3.8 \text{ gpm/ft}^2$$

PROBLEM 5.17

The raw wastewater diurnal flow variation and BOD concentration for a municipal wastewater treatment plant is shown in Figure 5.16. The BOD concentration represents an average value for four hour time period. Calculate the following:

(a) The minimum storage volume required for flow equalization.

(b) The BOD concentration and mass loading before and after flow equalization.

Solution

Figure 5.16 shows a graph of flow rate and BOD concentration versus time for a 24 hour period. The dotted line on the graph represents the average flow rate for each four hour period. As indicated in the problem statement, the BOD concentration shown is considered an average for the four hour time period. The BOD loading before equalization can be calculated from this data as shown in Table 5.11.

The required equalization volume is calculated graphically from a plot of cumulative volume versus time as shown in Figure 5.17. Lines are drawn tangent and parallel to the flow and the vertical distance between these two lines is the required minimum equalization volume of 1,200,000 gallons.

Using the equalization volume and Figure 5.16, the equalized BOD concentration for each four hour time interval is calculated. Figure 5.17 shows that the equalization basin is empty at approximately 8 a.m. Flow rate from 8 a.m. to 12 noon is 7.2 mgd or

$$\frac{7.2 \times 10^6 \text{ gallons}}{24 \text{ hrs}} \times 4 \text{ hrs} = 1.2 \times 10^6 \text{ gallons}$$

Influent BOD concentration is 210 mg/l from Figure 5.16.

BOD concentration in equalization basin is calculated as,

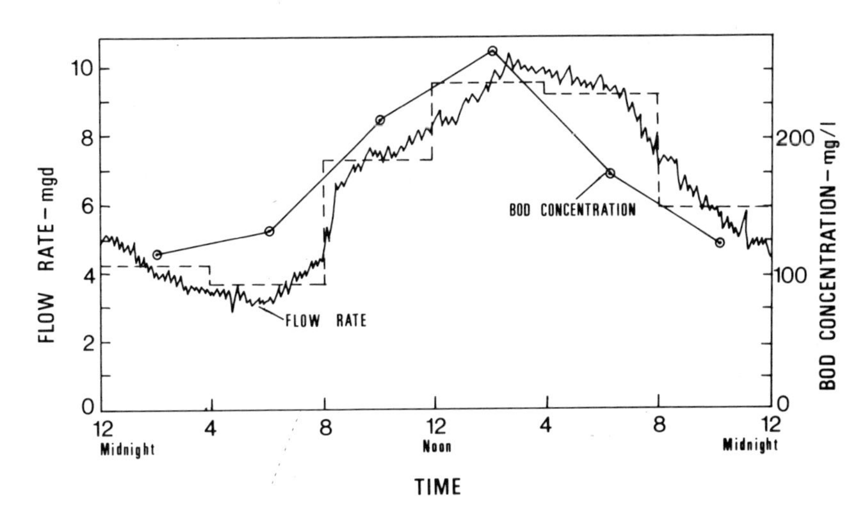

Figure 5.16. Diurnal flow pattern and BOD concentration for Problem 5.17.

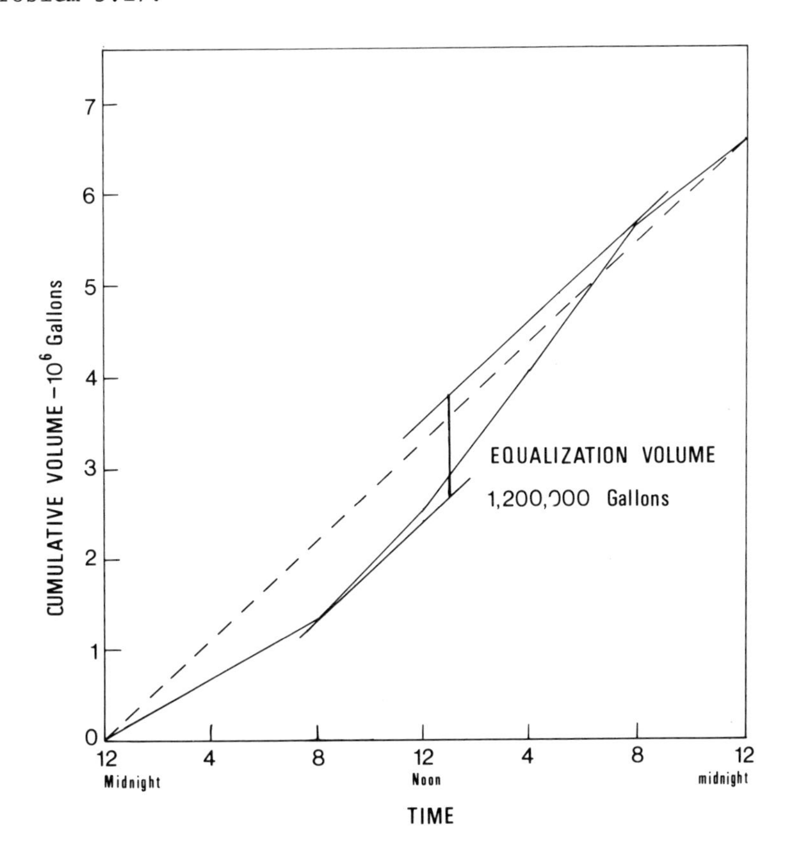

Figure 5.17. Calculation of equalization volume.

$$\frac{1{,}200{,}000 \text{ gallons} \times 210 \text{ mg/l}}{1{,}200{,}000 \text{ gallons}} = 210 \text{ mg/l}$$

Basin storage volume = 1.2 mg - 1.2 mg = 0

For 12 noon to 4 p.m. average flow rate is 9.5 mgd or 1,583,000 gallons.

Influent BOD concentration = 260 mg/l

BOD concentration (equalization basin) = 260 mg/l

Basin storage volume = 1,583,000 gallons - 1,200,000 gallons
= 383,000 gallons

For 4 p.m. to 8 p.m. average flow rate is 9.2 mgd or 1,533,000 gallons.

Influent BOD concentration = 170 mg/l

BOD concentration in equalization basin is calculated as,

$$\frac{(1{,}533{,}000 \times 170) + (383{,}000 \times 260)}{1{,}916{,}000} = 188 \text{ mg/l}$$

Basin storage volume = (1.533 mg - 1.2 mg) + 0.383 mg
= 0.716 mg

Similar calculations are made for remaining time periods and BOD concentrations are shown in Table 5.12. Since flow out of basin remains constant at 1,200,000 gallons, equalized BOD loading can be calculated. Results are shown in Table 5.12 and in Figure 5.18.

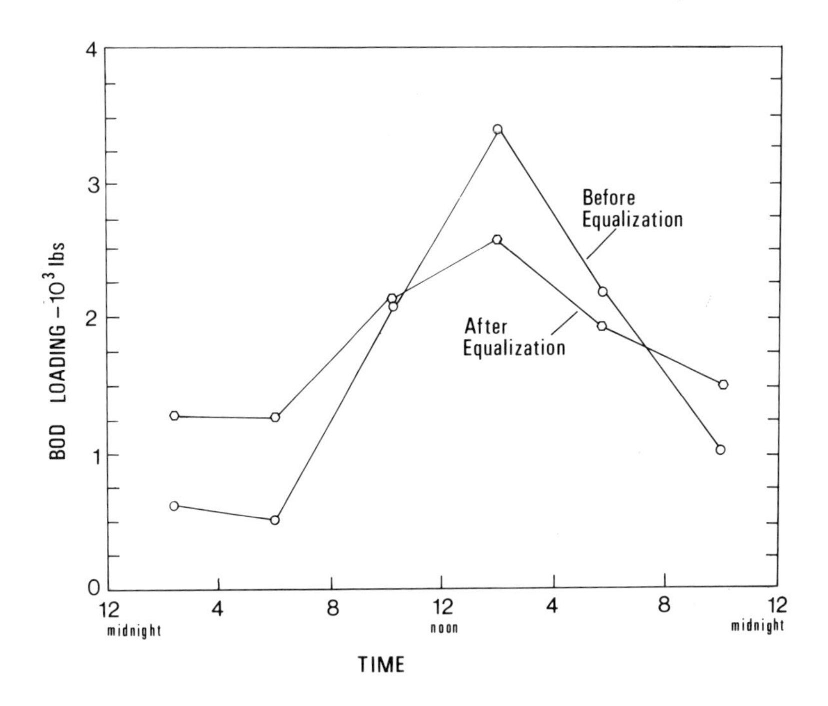

Figure 5.18. BOD loadings before and after equalization.

Table 5.11

BOD Concentrations and Loadings Before Equalization

Time	Flow Rate (mgd)	Flow (mg)	BOD (mg/l)	BOD (lbs)
12 Mid.-4 a.m.	4.25	0.708	115	680
4 a.m.-8 a.m.	3.50	0.583	130	635
8 a.m.-12 Noon	7.20	1.200	210	2100
12 Noon-4 p.m.	9.50	1.583	260	3430
4 p.m.-8 p.m.	9.20	1.533	170	2175
8 p.m.-12 Mid.	6.30	1.040	120	1040

Table 5.12

BOD Concentrations and Loadings After Equalization

Time	Flow (mg)	BOD (mg/1)	BOD (1bs)
12 Mid.-4 a.m.	1.2	129	1290
4 a.m.-8 a.m.	1.2	130	1300
8 a.m.-12 Noon	1.2	210	2100
12 Noon-4 p.m.	1.2	260	2600
4 p.m.-8 p.m.	1.2	188	1881
8 p.m.-12 Mid.	1.2	146	1460

References

1. U.S. Environmental Protection Agency, "Process Design Manual for Land Treatment of Municipal Wastewater," October, 1977.

2. Pennsylvania Department of Environmental Resources, "Spray Irrigation Manual," Publication No. 31, Bureau of Water Quality Management, 1972.

3. Atonie, Ronald L., Fixed Biological Surfaces - Wastewater Treatment, CRC Press, Inc., Cleveland, Ohio, 1976, pp. 47-75.

4. U.S. Environmental Protection Agency, "Process Design Manual for Nitrogen Control," October, 1975, pp. 4-42, 4-43.

5. "Denitrification by Anaerobic Filters and Ponds - Phase II," Robert S. Kerr Research Center, Ada, Oklahoma, Report No. 13030 ELY 06/71-14, June, 1971.

6. McCarty, P.L., L. Beck, and P. St. Amant, "Biological Denitrification of Wastewaters by Addition of Organic Materials," in Proceedings of the24th Industrial Waste Conference, May 6-8, 1969, Purdue University, Lafayette, Indiana, 1969.

7. Hutchins, Roy A., "New Method Simplifies Design of Activated Carbon Systems," Chemical Engineering, August 20, 1973, pp. 133-138.

8. DeWalle, Foppe B. and Edward S.K. Chian, "Biological Regeneration of Powdered Activated Carbon Added to Activated Sludge Units," Water Research, Vol. 11, No. 5, pp. 439-446, 1977.

Chapter 6

ENGINEERING ECONOMY

The following equations and terms are used in this Chapter to solve compound interest problems and problems in engineering economy. These interest formulas are from Principles of Engineering Economy[1] by Eugent L. Grant and W. Grant Ireson.

Definition of terms used:

P = present sum of money at time zero

n = number of interest periods

i = interest rate

S = sum of money at the end of n interest periods

R = end-of-period payments in a uniform series extending for n periods

A summary of the interest equations used in this chapter is given in Table 6.1.

The following discussion of several of the more common cost indexes is included as an introduction to this chapter. Cost indexes are useful to adjust costs for inflation and to make general or preliminary cost estimates. A special reprint of the Engineering New-Record Building Cost, Construction Cost, and Labor indexes by geographical location is included in the Appendix.

Table 6.1

INTEREST FORMULAS

Given	Find	Interest Factor Used	Abbreviation	Abbreviated Equation	Interest Formula
P	S	Single Payment Compound Amount	caf′	S = P (caf′)	$S = P\,(1+i)^n$
S	P	Single Payment Present Worth	pwf′	P = S (pwf′)	$P = S\,\frac{1}{(1+i)^n}$
S	R	Sinking Fund Factor	sff	R = S (sff)	$R = S\,\frac{i}{(1+i)^n-1}$
P	R	Capital Recovery Factor	crf	R = P (crf)	$R = P\,\frac{i\,(1+i)^n}{(1+i)^n-1}$
R	S	Uniform Series Compound Amount	caf	S = R (caf)	$S = R\,\frac{(1+i)^n-1}{i}$
R	P	Uniform Series Present Worth	pwf	P = R (pwf)	$P = R\,\frac{(1+i)^n-1}{i\,(1+i)^n}$

Cost Indexing Systems

Engineering cost estimates may be updated to current values using the following cost indexes:

Composite Cost Indexes

Engineering News-Record Construction Cost Index

Engineering News-Record Building Cost Index

Environmental Protection Agency Small City Conventional Treatment Index (SCCT)

Environmental Protection Agency Large City Advanced Treatment Index (LCAT)

Environmental Protection Agency Complete Urban Sewer System Index (CUSS)

Marshall and Swift Equipment Cost Index

Labor Indexes

Engineering News-Record Wage Labor Index

Engineering News-Record Skilled Labor Index

Present cost data can be obtained by applying the following cost index formula:

$$\text{Present Cost} = \text{Original Cost} \times \frac{\text{Present Index Value}}{\text{Original Index Value}}$$

A frequently used general cost index in the construction industry is the Engineering New-Record Construction Cost Index or Building Cost Index. The construction cost index is a 20-city composite index consisting of 200 hours of common labor, 2500 pounds of structural steel shapes, 1.128 tons of Portland Cement and 1,008 board feet of 2 x 4 lumber. The ENR building cost index consists of 68.38 hours of skilled labor and the same amount of material as the construction cost index. The ENR construction cost index is commonly used in the wastewater treatment field.

The Environmental Protection Agency-Sewage Treatment Plant index is a construction cost index frequently used to update costs for municipal wastewater treatment plants. The EPA-STP index has been updated to include three new cost indexes - the small city conventional treatment index, the large city advanced treatment index, and the complete urban sewer system index.[2] The base for the three indexes is the third quarter 1973. The SCCT index is based on a model 5 mgd activated sludge plant consisting of bar screens, grit chamber, primary clarification, conventional activated sludge, chlorination, gravity thickening, and vacuum filtration. The LCAT index is based on a model 50 mgd activated sludge plant. In addition to the unit processes for the 5 mgd plant, multi-media gravity filtration, lime clarification, and multiple hearth incineration is included in the LCAT index. The EPA complete urban sewer system index includes costs for collection and interceptor sewer and pump station and force main.

For more specialized applications such as construction of specific equipment items the Marshall and Swift index may be more applicable. The Marshall and Swift index is a composite industry equipment index and takes into consideration the cost of equipment plus costs for installation, fixtures, tools, and other minor equipment.

Both the Engineering News-Record skilled labor and common labor cost indexes are available for updating operation and maintenance costs. The ENR skilled labor index is usually preferred to calculate O & M costs.

PROBLEM 6.1

To finance construction of wastewater treatment plant, a municipality proposes to issue $500,000 worth of bonds that mature in 20 years and pay 8% interest on an annual basis. The bond issue will be financed by uniform annual deposits to a 20 year sinking fund that will earn 7% interest. Calculate the following:

(a) The annual interest payments to the bond holders.

(b) The required annual deposit to the sinking fund.

Solution

(a) Annual interest payment at 8%,

$$\$500{,}000 \times 0.08 = \$40{,}000$$

(b) Annual sinking fund deposit,

$R = S$ (sff-7%-20 yrs)

R = \$500,000 (0.02439)

R = \$12,195

The total annual cost to the municipality is \$52,195.

PROBLEM 6.2

Compare the total annual cost for a 30-inch and 36-inch diameter force main using the information shown in Table 6.2. Assume an interest rate of 8% and service life of 40 years.

Table 6.2

FORCE MAIN DATA

	30-inch	36-inch
Force main cost	\$90/LF	\$105/LF
Static pumping head	230 feet	230 feet
Overall pump efficiency	70%	70%
Operating period	8,760 hrs/yr	8,760 hrs/yr
Pipe length	10,850 feet	10,850 feet
Power cost	0.032¢/kwh	0.032¢/kwh
Flow rate	22.8 mgd	22.8 mgd

Solution

Economic analysis of force main costs requires comparison of pipe construction cost and operating cost. A smaller diameter pipe involves lower capital costs but higher energy costs because of the increased friction head for pumping. The friction head loss for each pipe is calculated for C = 100 and a flow rate of 22.8 mgd or 15,840 gpm.

30-inch: 0.81 ft/100 ft x 10,850 ft = 87.9 ft

36-inch: 0.33 ft/100 ft x 10,850 ft = 35.8 ft

Calculate total head for pumping (static and friction).

30-inch = 87.9 + 230 = 317.9 feet

36-inch = 35.8 + 230 = 265.8 feet

Calculate pumping costs,

$$Hp = \frac{Q\ H}{3960\ (efficiency)}$$

30-inch:

$$Hp = \frac{15{,}840\ gpm \times 317.9\ ft}{3960 \times 0.7}$$

Hp = 1817

$$1817\ Hp \times 0.746\ \frac{Kw}{Hp} \times 8760\ \frac{hr}{yr} = 11.87 \times 10^6\ kwh/yr$$

Pumping cost = $11.87 \times 10^6 \times \$0.032/kwh$
= \$379,900/year

36-inch:

$$Hp = \frac{15{,}840 \times 265.8\ ft}{3960 \times 0.70}$$

Hp = 1519

1519 Hp = 9.93×10^6 kwh/yr

Pumping cost = \$317,600/yr

Calculate annual cost of force main using a capital recovery factor of 0.08386 (8% interest rate for 40 years).

30-inch:

Construction cost = 10,850 ft x \$90/ft = \$976,500

Annual cost = \$976,500 x 0.08386 = \$81,890

36-inch:

Construction cost = \$1,139,250

Annual cost = \$95,540

Compare total annual cost (force main annual cost and annual pumping cost).

30-inch = \$461,790

36-inch = \$413,140

Therefore, the 36-inch pipe would be most economical on an annual cost basis.

PROBLEM 6.3

Three treatment alternatives are being considered for phosphorus removal at an activated sludge plant. Alternative I involves chemical addition using alum for phosphorus removal, incineration of the combined chemical and waste activated sludge, and disposal of the incinerator ash residue at a landfill. Alternative II is a proprietary treatment process for phosphorus removal requiring an increased capital expenditure but reduced chemical costs. The chemical sludge would be dewatered separately and disposed at the landfill. Waste activated sludge would be incinerated and the ash disposed at the landfill. Alternative III would use the same treatment process as Alternative II except both chemical and waste activated sludge would be incinerated and the ash disposed at the landfill. Table 6.3 gives the estimated cost of each alternative. Use a present worth analysis to determine the most cost-effective alternative. Assume a useful life of 20 years and an interest rate of 7%.

Table 6.3

Capital and Operating Costs for Three Treatment Alternatives for Problem 6.3.

	Alternative I	Alternative II	Alternative III
Capital Costs			
Construction	\$4,718,000	\$5,182,000	\$4,996,000
Salvage Value	377,000	265,000	250,000
Operating Costs			
Chemicals	138,700	81,700	72,700
Landfill Operation	23,600	47,800	24,800

Solution

Present worth value of each alternative can be calculated using the equation,

$PW = P - L\,(pwf') + AC(pwf)$

where P = initial capital investment

L = salvage value

AC = annual cost

PW = present worth value

For a 20 year service period and 7% interest rate,

pwf = 10.594

pwf′ = 0.2584

Alternative I

PW = $4,718,000 - $377,000(0.2584) + $162,300(10.594)

PW = $6,339,980

Alternative II

PW = $5,182,000 - $265,000(0.2584) + $129,500(10.594)

PW = $6,485,400

Alternative III

PW = $4,996,000 - $250,000(0.2584) + $97,000(10.594)

PW = $5,959,000

Alternative III is the most cost effective alternative based on a present worth analysis.

PROBLEM 6.4

In studying population and growth trends in a residential area it is estimated that water consumption will double over the next 20 years. The cost of expanding the existing water supply system will be compared with a phased program of expansion. Immediate development would cost \$420,000 with annual maintenance costs of \$40,000/year. A phased program would involve an initial investment of \$200,000 and an estimated expenditure of \$320,000 in 10 years. Annual maintenance cost is estimated at \$20,000/year for the first 10 years and at \$16,000/year following. Assuming a perpetual period of service for each system and an interest rate of 7%, evaluate the two alternatives on a capitalized cost basis.

Solution

Capitalized cost may be considered the same as a present worth analysis when an infinite period of time is considered.

For the immediate expansion program,

$$\text{Initial cost} = \$420{,}000$$

$$\text{Maintenance Cost} = \frac{\$40{,}000}{0.07} = \$571{,}400$$

$$\begin{aligned}\text{Capitalized cost} &= \$420{,}000 + \$571{,}400 \\ &= \$991{,}400\end{aligned}$$

For the phased expansion program,

$$\text{Initial cost} = \$200{,}000$$

$$\begin{aligned}\text{10-year cost} &= \$320{,}000\ (\text{pwf}'\text{-7\%-10 yrs}) \\ &= \$320{,}000\ (0.5083) \\ &= \$162{,}660\end{aligned}$$

Maintenance cost is calculated as \$16,000/year for an infinite period of time plus the present worth value of \$4,000/year (\$20,000-16,000) for an initial 10 year period.

$$\begin{aligned}\text{Maintenance cost} &= \frac{\$16{,}000}{0.07} + (\$20{,}000-\$16{,}000)(\text{pwf-7\%-10 yrs}) \\ &= \$228{,}600 + \$28{,}096 \\ &= \$256{,}696\end{aligned}$$

$$\begin{aligned}\text{Capitalized cost} &= \$200{,}000 + \$162{,}660 + \$256{,}696 \\ &= \$619{,}356\end{aligned}$$

The phased expansion program has the lower capitalized cost.

PROBLEM 6.5

Two types of sewer pipe are being evaluated for a construction project. Pipe A will cost $60/foot and have an estimated service life of 50 years. Annual cost for pumping and maintenance is estimated at $50,000 per year. Pipe B has an estimated service life of 60 years and an annual pumping and maintenance cost of $45,000 year. Approximately 9,000 feet of pipe is to be installed. Calculate how much more can be paid for Pipe B than Pipe A. Assume an interest rate of 8%.

Solution

Problem will be solved on an annual cost basis and the break-even value of Pipe B calculated.

For Pipe A,

Initial cost = $60/ft x 9000 ft = $540,000

Annual operating cost = $50,000

Capital Recovery Factor @ 8%, 50 years = 0.08174

For Pipe B,

Annual operating cost = $45,000/year

Capital Recovery Factor @ 8%, 60 years = 0.08080

Annual cost Pipe A = Annual cost Pipe B

$540,000 (0.08174) + $50,000 = B (0.08080) + $45,000

Cost pipe B = $608,200

Cost/ft = $67.58

The cost of pipe B should not exceed $67.58/foot

PROBLEM 6.6

An industry which discharges its wastewater to a municipal sewer system pays a sewer rate of 25 cents/1000 gallons based on metered water usage. The industry also pays a flow surcharge of 2.5 cents/1000 gallons, and a suspended solids surcharge based on the equation,

$$F = 0.017\ (SS-300)$$

where F = surcharge rate, cents/1000 gallons

SS = suspended solids concentration of waste

The present discharge of suspended solids is 1200 mg/l and water usage is 260,000 gallons/day. The industry operates 312 days per year.

Two options are being considered to reduce the rate for sewage service. The first option would be to install clarifier equipment to reduce the suspended solids concentration to 700 mg/l. Installed equipment cost would be $140,000. Annual operation and maintenance cost is estimated at $4000 per year. A second option would be to install the equipment and also make in-plant changes to reduce water usage by 30%. The additional cost would be $30,000. It is estimated that the suspended solids concentration in the discharge would be reduced to 500 mg/l.

The industry believes the additional expenditure is justified if the payback period is 15 years or less. Determine if either option is cost effective.

Solution

(a) Calculate current annual sewer charge before pretreatment is installed.

$$\text{Base rate} = \frac{\$0.25}{1000\ \text{gal}} \times 260{,}000\ \frac{\text{gal}}{\text{day}} \times 312\ \frac{\text{days}}{\text{year}}$$

$$= \$20{,}280/\text{year}$$

$$\text{Solids surcharge} = 0.017\ (1200-300) = \$15.3¢/1000\ \text{gal}$$

$$\frac{\$0.153}{1000\ \text{gal}} \times 260{,}000\ \frac{\text{gal}}{\text{day}} \times 312\ \frac{\text{days}}{\text{year}} = \$12{,}411/\text{year}$$

$$\text{Flow surcharge} = \frac{\$0.025}{1000\ \text{gal}} \times 260{,}000\ \frac{\text{gal}}{\text{day}} \times 312\ \frac{\text{days}}{\text{year}}$$

$$= \$2{,}028/\text{year}$$

Total Rate = $34,719/year

(b) Calculate sewer charge after installation of clarifier equipment.

Base rate = $20,280/year

Solids surcharge = 0.017 (700-300) = 6.8¢/1000 gal
= $5,516/year

Flow surcharge = $2,028/year

Total Rate = $27,824/year

Annual savings = sewer charge reduction-operating costs

Savings = ($34,719-$27,824)-$4000

= $2895/year

Equipment cost = $140,000

$$\text{Payback} = \frac{\$140{,}000}{\$2895/\text{year}} = 48\ \text{years}$$

(c) Calculate sewer charge after installation of clarifier equipment and 30% flow reduction. Assume base rate remains at 25¢/1000 gallons.

Flow rate = 182,000 gallons/day

$$\text{Base rate} = \frac{\$0.25}{1000\ \text{gal}} \times 182{,}000\ \frac{\text{gal}}{\text{day}} \times 312\ \frac{\text{days}}{\text{year}}$$

$$= \$14{,}196/\text{year}$$

Solids surcharge = 0.017 (500-300) = 3.4¢/1000 gal
= $1,931/year

Flow surcharge = $1,420/year

Total Rate = $17,547/year

Savings = ($34,719-$17,547)-$4000 = $13,172/year

$$\text{Equipment cost} = \$140{,}000 + \$30{,}000 = \$170{,}000$$

$$\text{Payback } \frac{\$170{,}000}{\$13{,}172/\text{year}} = 13 \text{ years}$$

Expenditure for installation of clarifier equipment plus reduction in flow rate would be justified since payback period is less than 15 years. Pretreatment required to meet any regulatory requirements should also be considered.

PROBLEM 6.7

Two 40 horsepower pumps are being considered for use as a standby pump to handle peak wastewater flows at a pumping station. Pump A costs $4,750 and has an overall rated efficiency of 84%. Pump B costs $3,550 and has an efficiency of 81%. Fixed charges for installation and maintenance amount to 12% of purchase cost and electricity cost is 5 cents per kilowatt-hour. How many hours of peak flow pumping each year would be necessary to justify purchase of the more expensive pump.

Solution

Let N = hours of peak flow pumping

Calculate electric costs for pumping N hours each year.

$$\text{Cost} = \frac{\$0.05}{\text{kw-hr}} \times 40 \text{ Hp} \times 0.7457 \frac{\text{kw}}{\text{Hp}} \times \frac{\text{N hrs}}{\text{efficiency}}$$

$$\text{Cost} = \$1.49 \frac{\text{N}}{\text{eff.}}$$

Calculate break-even value for N.

$$(\$4{,}750 \times 0.12) + \frac{1.49 \text{ N}}{0.84} = (\$3550 \times 0.12) + \frac{1.49 \text{ N}}{0.81}$$

$$\text{N} = 2190 \text{ hours/year}$$

PROBLEM 6.8

Land application using spray irrigation will be used for ultimate disposal of wastewater from a secondary treatment plant. Treatment plant effluent will be pumped a distance of 15,000 feet to the land application site. For a flow rate of 0.5 mgd and application rate of 2 inches/week, a total land area of 100 acres will be required. A storage reservoir with 45 days detention time will be provided.

Results of a preliminary cost estimate showing significant capital and O & M costs are summarized in Table 6.4. Assume that the total engineering, administrative, legal, and contingency costs are 12% of the capital cost. No revenue from sale of crops grown on the land will be realized. Calculate the annual cost using an interest rate of 7%. Assume the useful life of land is permanent, the useful life of the conveyance system is 50 years and the useful life of process equipment is 20 years.

Table 6.4

COSTS FOR LAND APPLICATION

Item	Construction Cost	Power	Labor	Maintenance
Land	$3,000/acre	-	-	-
Force Main	$18.75/LF	-	-	$15,000/yr
Effluent Pumping	$150,000	$20,000/yr	$8,000/yr	$1,000/yr
Storage Reservoir	$142,000	-	-	-
Spray Distribution System	$60,000	$840/yr	$14,000/yr	$750/yr
Monitoring Wells	$4,200	-	-	-

Solution

Calculate amortized cost for land. The salvage value of the land is calculated by assuming land value appreciates at a rate of 3% over a period of 20 years, the life of the process equipment.

caf′-3%-20 yrs = 1.806

Salvage value, L = $300,000 x 1.806 = $541,800

Amortized cost of land may be calculated from the relationship,

[P - L (pwf′)] crf

where P = present land value, $

L = salvage value, $

pwf′ = (pwf′-7%-20 yrs) = 0.2584

crf = (crf-7%-20 yrs) = 0.09439

Amortized Cost = [$300,000-$541,800(0.2584)] 0.09439

Amortized Cost = $15,100

Calculate amortized cost of wastewater conveyance system for a useful life of 50 years.

Force Main	$281,250
Pumping	150,000
Subtotal	$431,250
Eng., Legal, Admin., Contg. @12%	51,750
Total	$483,000

crf-7%-50 years = 0.07246

Amortized Cost = $483,000 x 0.07246 = $35,000

Calculate amortized cost for process equipment for a useful life of 20 years.

Storage	$142,000
Distribution	60,000
Monitoring Wells	4,200
Subtotal	$206,200
Eng., Legal, Admin., Contg. @ 12%	24,744
Total	$230,944

crf-7%-20 years = 0.09439

Amortized Cost = \$230,944 x 0.09439 = \$21,800

Total Operation and Maintenance Cost = \$59,600

Calculated total annual cost,

Land	\$ 15,100
Conveyance System	35,000
Process Equipment	21,800
Operation & Maintenance	59,600
Total	\$131,500

PROBLEM 6.9

Using the Engineering News-Record construction cost index and the EPA small city conventional treatment index, calculate the January 1978 cost for a wastewater treatment project that cost \$1,000,000 in January 1975.

Solution

(a) ENR Construction Cost Index (1967 Base = 100)

January 1975 = 196

January 1978 = 249

$$1978 \text{ Cost} = \$1{,}000{,}000 \times \frac{249}{196} = \$1{,}270{,}408$$

% Increase = 27%

(b) EPA Small City Conventional Treatment Index

1st Quarter 1975 = 109

1st Quarter 1978 = 132

$$1978 \text{ Cost} = \$1{,}000{,}000 \times \frac{132}{109} = \$1{,}211{,}009$$

% Increase = 21%

PROBLEM 6.10

A wastewater treatment plant will be upgraded using activated carbon columns. The cost of building a carbon regeneration system with multiple hearth furnace will be compared with purchasing carbon on a throwaway basis. Construction cost for a multiple hearth furnace is shown in Figure 6.1[3] and purchase, delivery, and placement cost of granular activated carbon is given in Figure 6.2.[4] Estimated costs for furnance operation and carbon regeneration design data are given in Table 6.5. Using this information determine the most cost effective alternative.

Table 6.5

DESIGN DATA AND OPERATING COSTS FOR CARBON REGENERATION

Design Data

Required carbon volume: 14,000 ft^3

Multiple hearth furnace loading: 40 lb/ft^2-day

Furnace downtime: 30%

Carbon regeneration rate: 150 lb/hr

Operating Cost

Labor: 7000 hours @ $10.00/hour

Electricity: 35,000 Kwh/year @ 3.5¢/Kwh

Fuel (natural gas): 3.5×10^6 scf/year @ 0.15¢/scf

Maintenance: $6,000/year

Solution

Calculation of carbon regeneration costs

At a furnace loading rate of 40 lb/sq. ft./day the required furnace size for regeneration is calculated as,

$$\frac{150 \text{ lb/hr} \times 24 \text{ hr/day}}{40 \text{ lb/ft}^2\text{/day}} = 90 \text{ ft}^2$$

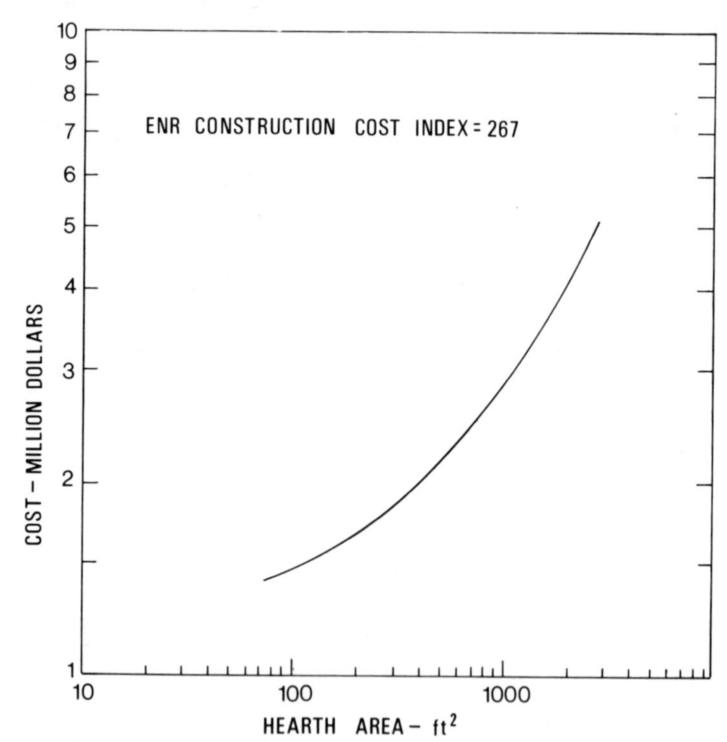

Figure 6.1. Multiple hearth furnace cost.

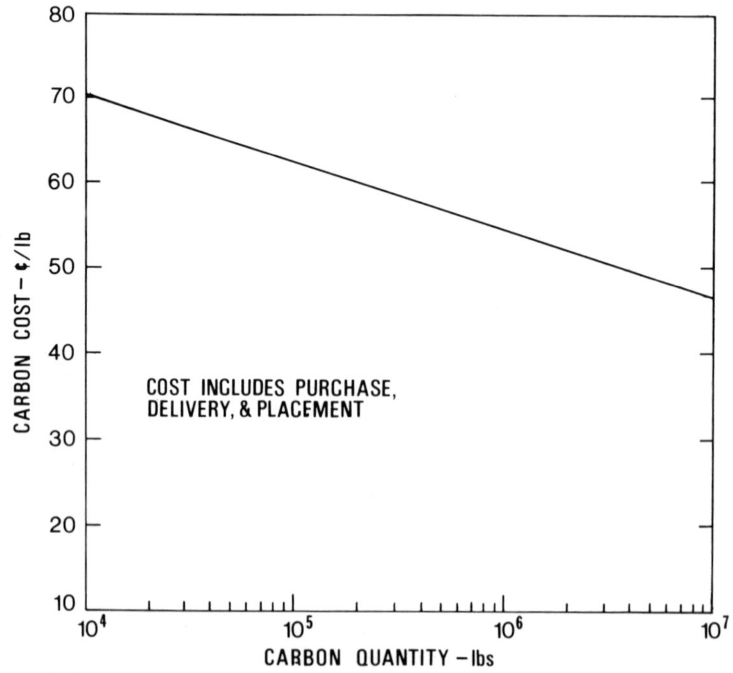

Figure 6.2. Cost of granular carbon.

Figure 6.1 has been updated to fourth quarter 1978 cost using the Engineering News Record (ENR) construction cost index. Cost includes a 25% allowance for contractor overhead and profit. It does not include allowance for engineering, legal, fiscal, administrative, and financing costs. From Figure 6.1, the construction cost for a multiple hearth carbon regeneration furnace with a 90 square foot area is $1,500,000.

The cost to initially charge the carbon columns is considered a capital cost and is calculated using a carbon density of 25 lbs/cu.ft.

Required carbon volume = 14,000 ft^3

Carbon weight = 14,000 ft^3 x 25 lb/ft^3 = 350,000 lbs

From Figure 6.2, the cost for purchase of 350,000 lbs carbon is approximately 54¢/pound.

350,000 lbs x $0.54/lb = $189,000

Carbon cost for regeneration operation must also be considered. Assuming a carbon loss of 10% through regeneration and transport and handling, make-up carbon cost is calculated based on a carbon feed rate of 150 lb/hr and 70% furnace operating time.

0.1 x 150 lb/hr x 24 x 365 x 0.7 = 92,000 lb/year

From Figure 6.2, carbon cost is taken as 65¢/lb.

92,000 lbs/year x $0.65 = $59,800/year

The total annual cost for carbon regeneration is calculated as $346,665 as shown in Table 6.6. An ammortization rate of 7% for a 20 year service period is used. No purchase of land is assumed to be required.

<u>Calculation of carbon costs on a throwaway basis</u>

Calculate the equivalent carbon replacement frequency for a carbon regeneration rate of 150 lb/hr and 70% furnace operation.

Carbon column exhaustion rate = 150 lb/hr x 0.7 = 105 lb/hr
= 2520 lb/day

Carbon column loading = 350,000 lbs

Table 6.6

COSTS FOR CARBON REGENERATION

Capital Costs		
Multiple Hearth Regeneration		$1,500,000
Initial Carbon Charge		189,000
	Construction Cost	$1,689,000
	Sitework @ 5%	84,450
		$1,773,450
Engineering @ 10%		177,345
	Subtotal	$1,950,795
Legal, Fiscal, and Administrative @ 3%		58,524
Interest During Construction - 8%		156,063
	TOTAL CAPITAL COSTS	$2,165,382
Operation and Maintenance Costs		
Labor		$ 70,000
Electricity		1,225
Fuel		5,250
Maintenance		6,000
Carbon Make-up		59,800
	TOTAL O & M COSTS	$ 142,275
Total Annual Cost		
Amortized Capital @ 7%, 20 years		$ 204,390
O & M Costs		142,275
	TOTAL ANNUAL COST	$ 346,665

$$\text{Carbon bed life} = \frac{350{,}000 \text{ lbs}}{2520 \text{ lb/day}} = 138 \text{ days}$$

Use a bed life of 4 months for calculations.

Carbon would be replaced 3 times/year.

Carbon required = 350,000 lbs x 3 = 1,050,000 lbs/year

Carbon cost = 53¢/lb

Carbon cost = 1,050,000 lbs/yr x \$0.53/lb = \$556,500/yr

Cost of carbon on a throwaway basis is approximately twice the cost of carbon regeneration. Site specific costs such as disposal of the waste carbon and carbon conveyance costs should also be considered in a more detailed cost analysis.

PROBLEM 6.11

The waste sludge from a 5 mgd wastewater treatment plant is thickened and dewatered using a vacuum filter. The daily quantity of thickened sludge is 10,000 pounds at 4% solids. An economic analysis will be made to determine if it will be more cost effective to install a vacuum filter to operate on an 8 hour, 16 hour, or 24 hour per day basis, seven days per week. Assuming the equation from Problem 4.7 is applicable for design of the vacuum filter unit, which operating schedule would be most cost effective?

Solution

From Problem 4.7, the cake yield for a vacuum filter can be calculated from the equation,

$$L = 3.85 \frac{C^{0.767}}{T^{0.656}}$$

where L = cake yield, $lbs/hr\text{-}ft^2$

T = cycle time, minutes/revolution

C = percent solids in feed

A cycle time of 5 minutes is assumed and cake yield is calculated as,

$$L = 3.88 \text{ lb/hr-ft}^2$$

The required filter area for an 8 hour operating day with a safety factor of 0.8 is calculated as,

Table 6.7

VACUUM FILTRATION COSTS

	Operating Period		
	8 hr/day	16 hr/day	24 hr/day
Capital Cost			
Vacuum Filter	$176,000	$110,000	$ 89,000
Sludge Pump	150,000	100,000	85,000
Subtotal	$326,000	$210,000	$174,000
Installation @ 20%	65,200	42,000	34,800
Total	$391,200	$252,000	$208,800
Operation and Maintenance			
Maintenance @ 4%	$ 13,040	$ 8,400	$ 6,960
Power @ 3.5¢/Kwh	4,200	5,250	4,620
Labor @ $8.00/hr	15,040	19,200	20,160
Ferric Chloride @ $500/ton	54,750	54,750	54,750
Total	$ 87,030	$ 87,600	$ 86,490
Amortized Capital Cost 7%, 20 years	$ 36,925	$ 23,785	$ 19,710
TOTAL ANNUAL COST	$123,955	$111,385	$106,200
Cost Per Ton Dry Solids	$67,90	$61.00	$58.19

$$\frac{10{,}000 \text{ lbs/day}}{8 \text{ hr/day} \times 3.88 \text{ lb/hr-ft}^2 \times 0.8} = 402 \text{ ft}^2$$

For 16 hour day, L = 201 ft^2

For 24 hour day, L = 134 ft^2

Table 6.7 shows a summary of capital and operating cost for each of the three operating periods to be evaluated. The vacuum filter cost is obtained from Figure 6.3[5], which has been updated to reflect fourth quarter 1978 equipment cost using a Marshall and Swift Index of 569. The vacuum filter equipment cost includes rotary drum vacuum filter, fabric filter cloth, pumps, drives, motors, vacuum receiver, chemical feed system, and electrical control panel. The costs for sludge pump and conveyor are not included. The cost for sludge pumps, power, and labor are obtained using Reference 3. Installation and maintenance is assumed to be 20% and 4%, respectively, of capital cost. Ferric chloride cost is based on a 6% (weight basis) feed rate. Capital cost is amortized at 7% over a 20 year period.

Economic analysis on an annual cost basis shows that operation 24 hours day would be the most cost effective alternative. Based on an annual sludge production rate of 1825 tons/year, cost would be approximately $58/ton. There is not a significant cost difference between a 16 hr/day and 24 hr/day operating period. Either a 16 hr/day or 20 hr/day operating period would most likely be considered from a practical operating standpoint. Other costs saved in a 16 hour to 24 hour operating period would be building cost and equipment costs such as conveyor that may be sized smaller with a longer operating day. Final recommendation should consider availability of required filter size and recommendation of equipment manufacturer.

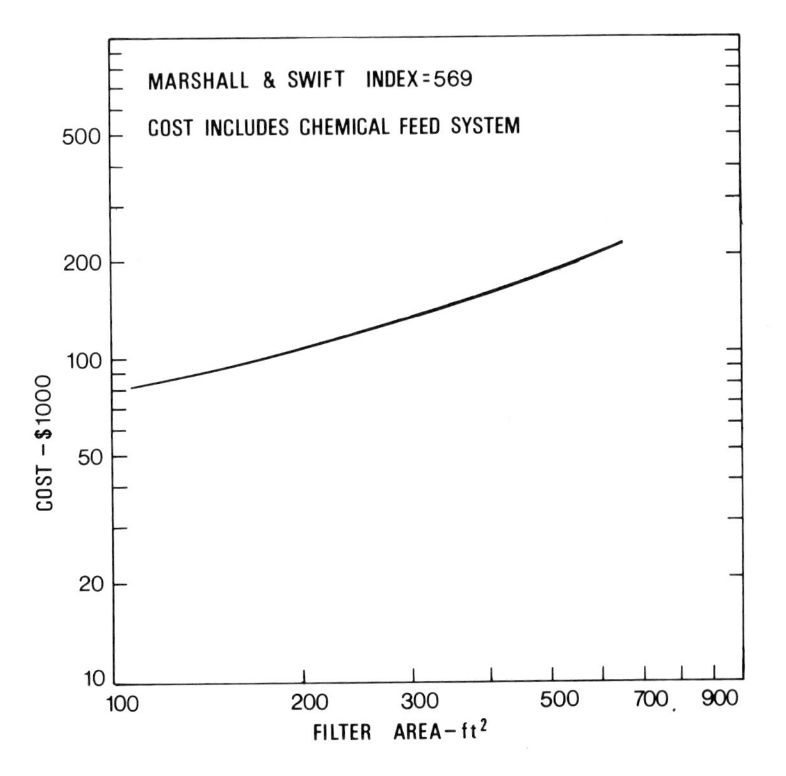

Figure 6.3 Vacuum filter cost. (Source: Sludge Dewatering Design Manual, Report No. 72, Ontario Ministry of the Environment.)

PROBLEM 6.12

Compare the present worth value of each alternative in Problem 6.11.

Solution

Present worth factor at 7% interest for 20 year service life = 10.594.

For 8 hour operating period,

Capital Cost = \$391,200

O & M Cost = \$87,030 x 10.594 = \$921,995

Total present worth = \$1,313,195

Present worth can also be calculated by dividing the annual cost by the capital recovery factor (crf), or multiplying the annual cost by the present worth factor (pwf).

$$\text{Present worth} = \frac{\text{Annual Cost}}{\text{crf}}$$

$$\text{Present worth} = \frac{\$123,955}{0.09439} = \$1,313,200$$

For 16 hour operating period,

$$\text{Present worth} = \frac{\$111,385}{0.09439} = \$1,180,050$$

For 24 hour operating period,

$$\text{Present worth} = \frac{\$106,200}{0.09439} = \$1,125,120$$

PROBLEM 6.13

Calculate the preliminary project cost for the wastewater treatment plant in Problem 3.2.

Table 6.8

PROJECT COSTS FOR 3.2 MGD ACTIVATED SLUDGE PLANT

Construction Costs	
Preliminary Treatment	$ 246,000
Influent Pumping	530,200
Primary Sedimentation	253,500
Activated Sludge Aeration	679,500
Final Clarifier	384,300
Chlorination	177,400
Anaerobic Digestion	448,900
Subtotal	$2,719,800
Engineering @ 7%	$ 190,400
Legal, Fiscal, and Administrative @ 3%	81,600
Contingencies @ 10%	271,980
TOTAL ESTIMATED PROJECT COST	$3,263,780

Solution

Preliminary project costs are summarized in Table 6.8. Construction costs for each unit process are obtained from Construction Costs for Municipal Wastewater Treatment Plants[6] and have been updated to second quarter 1978 dollars using the EPA small city conventional treatment index. Engineering fees are assumed to be 7% of construction costs, legal, fiscal, and administrative as 3%, and contingencies as 10%. Total estimated project cost is $3,263,780.

REFERENCES

1. Grant, Eugene L. and W. Grant Ireson, Principles of Engineering Economy, Ronald Press Company, New York, 1964.

2. Winklehaus, C., "EPA Switches to New Cost Indexes," Journal Water Pollution Control Federation, Vol. 48, No. 2, February, 1976, pp. 233-235.

3. Shuckrow, A.J. and G.L. Culp, "Appraisal of Powdered Activated Carbon Processes for Municipal Wastewater Treatment," EPA-600/2-77-156, September, 1977.

4. Gumerman, Robert C., Russell L. Culp, and Sigurd P. Hansen, "Estimating Costs for Water Treatment As A Function of Size and Treatment Efficiency," EPA-600/2-78-182, August, 1978.

5. Campbell, H.W., R.J. Rush, and R. Tew, Sludge Dewatering Design Manual, Research Report No. 72, Ontario Ministry of the Environment, Toronto, Ontario.

6. Dames and Moore, "Construction Costs for Municipal Wastewater Treatment Plants: 1973-1977," EPA-430/9-77-013, January, 1978.

APPENDIX 1

CONVERSION TABLE

To Convert From	To	Multiply By
	AREA	
acre	sq ft	43,560
acre	sq meters	4047
acre	sq miles	1.562×10^{-3}
sq in	sq cm	6.452
sq ft	sq meters	0.0929
sq ft	acres	2.2956×10^{-5}
sq cm	sq in	0.155
sq cm	sq ft	0.00108
sq meter	sq ft	10.76
sq km	sq mile	0.3861
sq km	sq ft	1.076×10^{7}
sq km	acres	247.1
	DENSITY	
gm/cu cm	kg/cu meter	1000
gm/cu cm	kg/liter	1
gm/cu cm	lbs/cu ft	62.43
gm/cu cm	lbs/gal (U.S.)	8.345
gm/cu cm	lbs/gal (Br.)	10.022
kg/cu meter	gm/cu cm	0.001
kg/cu meter	kg/liter	0.001
kg/cu meter	lbs/cu ft	0.0624
lbs/cu in	Kg/cu meter	2.768×10^{4}
lbs/cu ft	Kg/liter	0.01602
lbs/cu ft	kg/cu meter	16.018
lbs/cu ft	lbs/gal (U.S.)	0.1337
lbs/gal (U.S.)	kg/liter	0.1198
lbs/gal (U.S.)	lbs/cu ft	7.48
	FLOW RATE	
cu ft/sec	gal/min	448.8
cu ft/sec	liters/sec	28.32
cu ft/sec	mil gal/day	0.6462
cu m/hr	gal/min	4.4

To Convert From	To	Multiply By
	FLOW RATE	
gal/min	cu ft/sec	0.00223
gal/min	cu ft/min	0.1337
gal/min	liters/sec	0.0631
liters/sec	gal/min	15.85
liters/sec	cu ft/min	2.119
mil gal/day	cu ft/sec	1.547
	LENGTH	
feet	cm	30.48
inch	cm	2.54
km	mile	0.6214
km	feet	3280.8
meter	inch	39.37
meter	feet	3.281
meter	mile	6.21×10^{-4}
mile	feet	5280
mile	km	1.6093
	MASS	
gram	lb	2.205×10^{-3}
gram	ounce	0.03527
gram	grain	15.43
kg	lb	2.205
lb	grain	7000
lb	gram	453.6
lb	ounce	16
	POWER	
Btu	cal	252
Btu	ft-lb	778
Btu	hp-hr	3.93×10^{-4}
Btu	Kw-hr	2.93×10^{-4}
hp	Btu/min	42.44
hp	ft-lb/sec	550
hp	kw	0.7457
hp-hr	Btu	2545
hp-hr	Kw-hr	0.7457
hp-hr	ft-lb	1.98×10^{6}

To Convert From	To	Multiply By
	POWER	
kw	Btu/min	56.9
kw	ft-lb/sec	737.6
kw	hp	1.341
kw-hr	Btu	3413
kw-hr	ft-lb	2.66×10^6
kw-hr	hp-hr	1.341
	PRESSURE	
atm	mm Hg	760
atm	in Hg	29.92
atm	ft water	33.93
atm	meter water	10.33
atm	lbs/sq in	14.7
atm	Kg/sq meter	1.033×10^4
atm	newton/sq m	1.013×10^5
ft water	atm	0.02947
ft water	mm Hg	22.398
ft water	in Hg	0.8818
ft water	lb/sq in	0.433
ft water	lb/sq ft	62.43
ft water	kg/sq meter	304.8
ft water	newton/sq m	2989
in water	in Hg	0.0735
in water	mm Hg	1.8665
in Hg	ft water	1.133
in Hg	lb/sq in	0.49116
in Hg	mm Hg	25.40
mm Hg	in water	0.5358
mm Hg	in Hg	0.03937
mm Hg	lb/sq in	0.01934
mm Hg	Kg/sq meter	13.595
lb/sq in	atm	0.06805
lb/sq in	ft water	2.304
lb/sq in	in Hg	2.036
lb/sq in	kg/sq cm	0.0703
lb/sq in	newton/sq m	6895
	TEMPERATURE	
Fahrenheit	Centigrade	5/9 (F-32)
Centigrade	Fahrenheit	1.8 C + 32
Kelvin	Centigrade	C + 273.16
Rankine	Fahrenheit	F + 459.69

To Convert From	To	Multiply By
	VISCOSITY (ABSOLUTE)	
centipoise	lb_f sec/sq ft	2.09×10^{-5}
centipoise	lb_m/ft sec	6.72×10^{-4}
centipoise	gm/cm sec	0.01
	VISCOSITY (KINEMATIC)	
centistokes	sq cm/sec	0.01
centistokes	sq ft/sec	1.076×10^{-5}
	VELOCITY	
ft/sec	cm/sec	30.48
ft/sec	meter/sec	0.3048
ft/sec	km/hr	1.097
ft/min	cm/sec	0.508
cm/sec	ft/sec	0.03281
cm/sec	meter/min	0.60
meter/sec	ft/sec	3.281
	VOLUME	
cu ft	cu meter	0.02832
cu ft	gal (U.S.)	7.481
cu ft	gal (Br.)	6.229
cu ft	liter	28.316
cu meter	cu ft	35.314
cu meter	gal (U.S.)	264.17
cu meter	gal (Br.)	220.1
cu meter	liter	1000
gal	cu ft	0.1337
gal	ounce	128
gal	pint	8
gal	quart	4
gal (U.S.)	gal (Br.)	0.833
gal	liter	3.785
liter	cu meter	0.001
liter	cu in	61.025
liter	cu ft	0.03532
liter	fl ounce	33.81
liter	fl pint	2.113
liter	fl quart	1.057
liter	gal (U.S.)	0.2642
liter	gal (Br.)	0.22

APPENDIX 2

MOLECULAR AND EQUIVALENT WEIGHTS

Name	Formula	Molecular Weight	Equivalent Weight
Aluminum Sulfate	$Al_2(SO_4)_3 \cdot 18H_2O$	666.4	111.0
Aluminum Sulfate	$Al_2(SO_4)_3$	342.1	57.0
Aluminum Hydroxide	$Al(OH)_3$	78.0	26.0
Bicarbonate	HCO_3^-	61.0	61.0
Calcium	Ca	40.1	20.0
Calcium Bicarbonate	$Ca(HCO_3)_2$	162.1	81.0
Calcium Carbonate	$CaCO_3$	100.0	50.0
Calcium Hydroxide	$Ca(OH)_2$	74.1	37.0
Calcium Oxide	CaO	56.1	28.0
Calcium Sulfate	$CaSO_4$	136.2	68.1
Carbon Dioxide	CO_2	44.0	22.0
Ferrous Sulfate	$FeSO_4 \cdot 7H_2O$	278.0	139.0
Ferric Chloride	$FeCl_3$	162.2	54.1
Ferric Sulfate	$Fe_2(SO_4)_3$	400.0	66.7
Hydrogen Sulfide	H_2S	34.1	17.0
Hydrochloric Acid	HCl	36.5	36.5
Magnesium	Mg	40.3	20.1
Nitric Acid	HNO_3	63.0	63.0
Sodium	Na	23.0	23.0
Sodium Bicarbonate	$NaHCO_3$	84.0	84.0
Sodium Carbonate	Na_2CO_3	106.0	53.0
Sodium Hydroxide	NaOH	40.0	40.0
Sulfuric Acid	H_2SO_4	98.1	49.0
Sulfate	SO_4^{-2}	96.0	48.0

APPENDIX 3

ENGINEERING NEWS-RECORD BUILDING COST INDEX 1913-1978

1913	100	**1920**	207	**1927**	186	**1934**	167	**1941**	211	**1948**	345	**1955**	469
1914	92	**1921**	166	**1928**	188	**1935**	166	**1942**	222	**1949**	352	**1956**	491
1915	95	**1922**	155	**1929**	191	**1936**	172	**1943**	229	**1950**	375	**1957**	509
1916	131	**1923**	186	**1930**	185	**1937**	196	**1944**	235	**1951**	401	**1958**	525
1917	167	**1924**	186	**1931**	168	**1938**	197	**1945**	239	**1952**	416	**1959**	548
1918	159	**1925**	183	**1932**	141	**1939**	197	**1946**	262	**1953**	431	**1960**	559
1919	159	**1926**	185	**1933**	148	**1940**	203	**1947**	313	**1954**	446	**1961**	568

1913 = 100						**Monthly**							
	Jan.	**Feb.**	**Mar.**	**Apr.**	**May**	**June**	**July**	**Aug.**	**Sept.**	**Oct.**	**Nov.**	**Dec.**	**Annual Average**
1962	571	573	575	576	579	580	583	586	586	586	584	584	**580**
1963	584	585	586	586	588	590	596	602	602	604	602	603	**594**
1964	604	604	606	607	609	612	615	616	617	617	617	617	**612**
1965	616	621	622	621	621	626	628	630	633	634	633	634	**627**
1966	635	641	643	649	652	656	653	655	656	655	655	655	**650**
1967	656	657	659	660	666	671	673	678	681	684	685	687	**672**
1968	692	695	698	701	710	718	721	729	741	747	747	755	**721**
1969	764	770	780	790	791	798	792	799	796	797	801	802	**790**
1970	802	801	802	813	827	834	848	851	857	862	866	866	**836**
1971	875	877	905	913	933	946	959	970	996	997	1001	1005	**948**
1972	1011	1016	1022	1027	1039	1047	1053	1057	1067	1070	1082	1090	**1048**
1973	1102	1114	1123	1135	1140	1138	1137	1144	1150	1156	1155	1158	**1138**
1974	1156	1154	1155	1177	1177	1199	1233	1240	1238	1246	1239	1240	**1204**
1975	1242	1265	1265	1269	1287	1307	1317	1330	1333	1351	1349	1354	**1306**
1976	1362	1370	1378	1391	1398	1416	1425	1455	1467	1476	1479	1484	**1425**
1977	1489	1499	1504	1506	1507	1521	1539	1554	1587	1618	1604	1607	**1545**
1978	1609	1617	1620	1621	1652	1663	1696	1705	1720	1721	1732	1734	**1674**
1979	1740	1740	1750										

ENGINEERING NEWS-RECORD CONSTRUCTION COST INDEX
1913-1978

Year	Index	Year	Index	Year	Index	Year	Index	Year	Index	Year	Index	Year	Index
1906	95	**1914**	89	**1922**	174	**1930**	203	**1938**	236	**1946**	346	**1954**	628
1907	101	**1915**	93	**1923**	214	**1931**	181	**1939**	236	**1947**	413	**1955**	660
1908	97	**1916**	130	**1924**	215	**1932**	157	**1940**	242	**1948**	461	**1956**	692
1909	91	**1917**	181	**1925**	207	**1933**	170	**1941**	258	**1949**	477	**1957**	724
1910	96	**1918**	189	**1926**	208	**1934**	198	**1942**	276	**1950**	510	**1958**	759
1911	93	**1919**	198	**1927**	206	**1935**	196	**1943**	290	**1951**	543	**1959**	797
1912	91	**1920**	251	**1928**	207	**1936**	206	**1944**	299	**1952**	569	**1960**	824
1913	100	**1921**	202	**1929**	207	**1937**	235	**1945**	308	**1953**	600	**1961**	847

1913 = 100

	Monthly												
	Jan.	**Feb.**	**Mar.**	**Apr.**	**May**	**June**	**July**	**Aug.**	**Sept.**	**Oct.**	**Nov.**	**Dec.**	**Annual Average**
1962	855	858	861	863	872	873	877	881	881	880	880	880	**872**
1963	883	883	884	885	894	899	909	914	914	916	914	915	**901**
1964	918	920	922	926	930	935	945	948	947	948	948	948	**936**
1965	948	957	958	957	958	969	977	984	986	986	986	988	**971**
1966	988	997	998	1006	1014	1029	1031	1033	1034	1032	1033	1034	**1019**
1967	1039	1041	1043	1044	1059	1068	1078	1089	1092	1096	1097	1098	**1070**
1968	1107	1114	1117	1124	1142	1154	1158	1171	1186	1190	1191	1201	**1155**
1969	1216	1229	1238	1249	1258	1270	1283	1292	1285	1299	1305	1305	**1269**
1970	1309	1311	1314	1329	1351	1375	1414	1418	1421	1434	1445	1445	**1385**
1971	1465	1467	1496	1513	1551	1589	1618	1629	1654	1657	1665	1672	**1581**
1972	1686	1691	1697	1707	1735	1761	1772	1777	1786	1794	1808	1816	**1753**
1973	1838	1850	1859	1874	1880	1896	1901	1902	1929	1933	1935	1939	**1895**
1974	1940	1940	1940	1961	1961	1993	2040	2076	2089	2100	2094	2101	**2020**
1975	2103	2128	2128	2135	2164	2205	2248	2274	2275	2293	2292	2297	**2212**
1976	2305	2314	2322	2327	2357	2410	2414	2445	2465	2478	2486	2490	**2401**
1977	2494	2505	2513	2514	2515	2541	2579	2611	2644	2675	2659	2669	**2577**
1978	2672	2681	2693	2698	2733	2753	2821	2829	2851	2851	2861	2869	**2776**
1979	2872	2877	2886										

Reprinted with permission from the March 22, 1979 issue of Engineering News-Record magazine.

ENGINEERING NEWS-RECORD COST INDEXES IN 22 CITIES

BCI-Building Cost Index, CCI-Construction Cost Index, MCC-Materials Cost Component Index, SLI-Skilled Labor Index, CLI-Common Labor Index.

1967 = 100	ATLANTA					BALTIMORE					BIRMINGHAM					BOSTON					CHICAGO				
	BCI	CCI	MCC	SLI	CLI	BCI	CCI	MCC	SLI	CLI	BCI	CCI	MCC	SLI	CLI	BCI	CCI	MCC	SLI	CLI	BCI	CCI	MCC	SLI	CLI
1954 Dec. ...	67	58	77	59	49	71	66	84	60	60	70	61	77	64	52	66	62	81	54	55	63	61	70	57	57
1955 Dec. ...	70	61	81	61	51	74	67	89	62	60	74	64	81	67	55	71	65	88	57	57	66	63	74	59	59
1956 Dec. ...	74	64	85	64	54	78	68	92	66	60	76	66	84	68	57	74	69	91	61	60	68	66	78	60	61
1957 Dec. ...	76	67	88	66	57	81	76	95	70	71	80	71	87	72	63	76	71	90	65	63	71	69	79	64	66
1958 Dec. ...	79	72	90	69	63	85	80	98	74	76	82	74	90	75	66	79	74	92	68	67	76	71	86	67	66
1959 Dec. ...	83	77	95	73	67	87	84	100	77	82	86	78	94	78	70	81	76	94	70	69	78	76	89	70	71
1960 Dec. ...	84	80	93	76	74	87	85	98	79	84	88	79	95	81	70	83	78	96	73	71	79	76	87	73	71
1961 Dec. ...	84	80	91	78	74	87	85	94	81	85	90	84	97	84	77	87	82	100	76	75	80	78	85	76	76
1962 Dec. ...	87	84	94	82	79	87	87	92	84	90	91	84	97	85	78	88	82	99	79	76	82	79	86	79	76
1963 Dec. ...	90	88	97	84	83	90	87	94	87	90	94	89	99	89	83	92	87	100	85	82	87	83	91	83	80
1964 Dec. ...	94	91	96	89	89	91	90	93	90	94	96	91	99	94	87	92	91	97	88	89	89	87	91	88	85
1965 Dec. ...	99	97	107	93	92	93	92	92	93	97	95	93	95	96	92	92	92	93	91	91	93	92	96	91	90
1966 Dec. ...	98	98	100	97	98	99	99	101	96	103	99	97	100	98	96	97	98	99	97	98	97	96	99	97	95
1967 Dec. ...	103	103	103	103	103	102	101	100	103	108	103	106	101	104	108	102	102	101	104	102	102	103	100	102	103
1968 Dec. ...	115	113	119	111	110	111	111	114	108	116	111	112	112	111	113	112	109	114	110	107	109	110	107	111	112
1969 Dec. ...	120	120	119	121	120	119	115	117	120	121	115	119	111	118	123	119	121	112	124	124	117	117	109	124	120
1970 Dec. ...	127	135	117	136	144	130	137	113	144	158	117	131	111	123	142	128	132	115	138	138	125	131	107	141	141
1971 Dec. ...	149	153	143	155	159	158	171	145	168	194	142	148	135	149	156	147	159	141	152	165	143	152	123	159	164
1972 Dec. ...	166	174	162	170	181	167	184	155	177	210	156	164	145	167	174	163	174	155	169	182	153	163	131	172	175
1973 Dec. ...	173	189	168	177	199	183	201	180	186	224	165	170	157	173	177	175	182	171	178	186	165	173	144	181	184
1974 Dec. ...	181	196	175	186	207	181	207	171	191	238	180	187	166	193	199	187	199	183	191	205	176	183	155	193	194
1975 Dec. ...	194	212	197	192	220	203	226	197	208	255	195	201	186	205	210	204	223	201	205	232	188	194	169	203	204
1976 Dec. ...	216	232	230	205	232	225	253	226	224	283	219	227	222	217	231	220	222	227	215	220	206	214	194	215	221
1977 Mar. ...	222	239	238	210	240	227	254	228	227	283	219	228	223	217	231	221	222	229	215	220	206	214	194	215	221
June ...	221	237	233	211	240	229	255	231	227	283	218	227	220	217	231	221	222	229	215	220	210	214	195	223	221
Sept. ...	231	245	251	214	243	240	264	255	228	285	238	255	249	229	260	234	237	243	227	235	223	230	218	227	235
Dec. ...	231	245	251	214	243	238	262	249	228	285	234	252	239	229	260	245	244	266	229	235	226	232	226	227	235
1978 Mar. ...	234	248	258	214	243	241	264	256	228	285	239	253	243	236	260	238	242	247	230	240	228	233	228	227	235
June ...	244	256	268	224	250	248	269	269	230	285	247	259	260	236	260	242	245	259	229	240	234	233	227	239	235
Sept. ...	248	265	278	224	259	250	270	274	230	285	256	272	262	251	278	250	250	268	236	243	242	249	245	239	250
Dec. ...	251	267	283	224	259	256	272	279	237	285	258	274	267	251	278	250	256	269	236	251	240	247	240	239	250
1979 Mar. ...	249	267	272	230	264	260	275	288	237	285	258	273	265	252	278	255	257	275	240	251	243	249	247	239	250

ENGINEERING NEWS-RECORD COST INDEXES IN 22 CITIES

1967 = 100	CINCINNATI					CLEVELAND					DALLAS				
	BCI	CCI	MCC	SLI	CLI	BCI	CCI	MCC	SLI	CLI	BCI	CCI	MCC	SLI	CLI
1954 Dec. ...	70	60	78	61	53	66	60	78	56	55	67	61	64	66	58
1955 Dec. ...	73	63	82	63	55	69	63	84	57	56	70	65	69	68	60
1956 Dec. ...	77	67	87	66	59	71	65	87	60	59	74	68	73	71	62
1957 Dec. ...	79	70	88	69	63	74	69	88	63	63	75	69	73	73	65
1958 Dec. ...	81	73	89	71	66	75	70	89	65	65	78	73	76	75	68
1959 Dec. ...	84	77	93	74	69	78	73	92	67	68	81	76	80	78	72
1960 Dec. ...	85	79	92	77	73	79	75	92	69	71	87	81	89	80	74
1961 Dec. ...	87	81	93	78	76	80	78	92	72	74	88	83	88	82	78
1962 Dec. ...	88	82	94	80	76	81	80	91	74	77	90	84	91	85	78
1963 Dec. ...	91	87	96	84	82	84	83	95	77	80	93	86	92	88	80
1964 Dec. ...	93	89	94	88	86	87	86	95	81	84	94	89	93	91	84
1965 Dec. ...	100	95	106	91	90	92	90	98	88	88	97	93	94	94	90
1966 Dec. ...	99	97	99	96	95	96	95	97	95	95	102	100	100	98	97
1967 Dec. ...	106	107	101	106	107	105	106	102	108	108	104	105	101	101	104
1968 Dec. ...	121	126	117	120	128	119	119	117	121	121	114	113	109	112	112
1969 Dec. ...	131	144	119	137	151	126	127	118	133	132	121	135	109	126	147
1970 Dec. ...	140	151	119	152	161	137	140	120	150	148	126	146	107	137	165
1971 Dec. ...	168	186	143	182	199	156	157	142	166	163	143	162	130	148	176
1972 Dec. ...	175	192	151	188	205	165	169	152	175	175	160	188	144	166	208
1973 Dec. ...	190	206	175	195	214	172	177	160	182	183	170	199	152	178	221
1974 Dec. ...	207	226	196	210	234	187	191	178	194	197	177	211	156	187	237
1975 Dec. ...	217	241	199	225	254	201	206	191	208	212	192	223	176	197	244
1976 Dec. ...	232	250	221	234	257	218	223	212	223	229	216	238	202	217	252
1977 Mar. ...	236	252	228	235	257	218	223	211	223	229	219	240	209	217	252
June ...	242	255	241	235	257	227	230	212	237	238	220	242	200	226	260
Sept. ...	254	266	248	251	269	228	237	215	237	246	238	256	236	226	260
Dec. ...	258	268	257	251	269	231	239	223	237	246	235	254	230	226	260
1978 Mar. ...	259	268	257	252	269	232	239	226	237	246	241	257	238	231	260
June	265	278	271	252	277	237	249	237	237	255	245	260	245	231	260
Sept.	277	288	277	268	288	246	256	239	252	264	254	264	255	239	260
Dec.	282	291	288	268	228	254	261	259	252	264	258	264	255	245	260
1979 Mar.	284	291	290	270	288	252	260	253	252	264	258	264	256	245	260

1967 = 100	DENVER					DETROIT					KANSAS CITY				
	BCI	CCI	MCC	SLI	CLI	BCI	CCI	MCC	SLI	CLI	BCI	CCI	MCC	SLI	CLI
1954 Dec. ...	71	65	87	60	57	63	58	73	55	53	72	64	81	63	57
1955 Dec. ...	74	67	90	61	58	67	62	80	57	55	76	68	89	65	59
1956 Dec. ...	79	71	100	63	61	71	65	86	60	57	78	69	90	66	60
1957 Dec. ...	80	73	98	66	64	74	66	88	62	59	80	72	91	69	63
1958 Dec. ...	81	75	100	67	66	78	71	94	65	63	82	74	93	72	66
1959 Dec. ...	83	76	100	71	68	81	77	97	67	69	85	77	96	75	69
1960 Dec. ...	84	77	99	74	70	82	77	97	70	70	85	79	93	78	72
1961 Dec. ...	86	80	97	78	74	83	79	97	72	72	86	80	91	81	76
1962 Dec. ...	87	82	96	80	78	85	81	98	75	75	88	83	92	83	79
1963 Dec. ...	89	85	98	83	82	88	83	100	78	77	90	86	94	86	83
1964 Dec. ...	90	88	98	86	86	87	85	93	82	82	92	89	95	90	87
1965 Dec. ...	91	90	97	88	90	92	90	96	89	87	95	93	97	93	92
1966 Dec. ...	94	95	99	92	95	97	96	98	96	95	98	98	100	97	97
1967 Dec. ...	98	99	100	98	102	103	103	100	104	104	102	102	103	101	102
1968 Dec. ...	113	112	127	103	109	112	112	105	117	114	108	108	109	107	107
1969 Dec. ...	114	115	108	121	122	126	122	117	133	124	116	130	115	117	137
1970 Dec. ...	124	123	109	139	133	133	136	106	153	147	153	168	158	148	171
1971 Dec. ...	146	138	133	159	145	150	156	125	169	167	153	174	133	171	191
1972 Dec. ...	160	151	151	170	155	160	168	136	178	178	168	191	148	186	210
1973 Dec. ...	167	159	154	180	167	169	178	147	186	189	180	205	162	197	223
1974 Dec. ...	179	171	167	193	178	183	191	160	200	202	195	223	176	213	243
1975 Dec. ...	197	193	189	207	201	198	206	174	215	217	214	243	202	224	260
1976 Dec. ...	217	217	214	222	225	215	222	197	228	230	233	260	227	239	274
1977 Mar. ...	224	222	230	222	225	217	223	201	228	230	237	262	235	239	274
June ...	235	237	229	243	248	218	223	202	228	230	240	267	235	244	280
Sept. ...	238	239	236	243	248	229	238	212	241	246	244	274	239	249	289
Dec. ...	241	241	243	243	248	231	239	218	241	246	256	280	257	254	289
1978 Mar. ...	242	241	244	243	248	232	240	219	241	246	259	280	258	261	289
June	246	244	255	243	248	237	243	231	241	246	269	293	268	271	303
Sept.	254	259	261	252	266	250	256	232	264	263	271	294	271	271	303
Dec.	259	262	272	252	266	253	257	238	264	263	271	294	271	271	303
1979 Mar.	259	262	272	253	266	251	256	234	264	263	272	295	273	271	303

Reprinted with permission from the March 22, 1979 issue of Engineering News-Record magazine.

ENGINEERING NEWS-RECORD COST INDEXES IN 22 CITIES

	LOS ANGELES					MINNEAPOLIS					NEW ORLEANS					NEW YORK					PHILADELPHIA				
	BCI	CCI	MCC	SLI	CLI	BCI	CCI	MCC	SLI	CLI	BCI	CCI	MCC	SLI	CLI	BCI	CCI	MCC	SLI	CLI	BCI	CCI	MCC	SLI	CLI
1954 Dec. ...	66	57	86	54	48	71	61	92	59	51	71	63	84	60	52	66	54	93	52	43	77	67	98	62	56
1955 Dec. ...	70	60	92	57	51	73	63	93	60	53	74	65	88	62	54	69	57	97	53	46	80	70	103	63	58
1956 Dec. ...	72	62	93	59	53	76	66	99	62	56	77	67	89	65	56	72	61	100	56	50	82	73	108	63	60
1957 Dec. ...	75	66	95	62	58	78	69	90	65	59	79	71	92	68	60	74	64	101	60	54	83	76	107	67	64
1958 Dec. ...	77	70	97	65	62	81	72	102	68	63	81	74	94	70	64	77	68	104	62	58	85	77	105	70	67
1959 Dec. ...	80	74	99	70	67	83	75	102	71	67	84	77	95	74	67	81	73	105	67	64	86	77	105	72	67
1960 Dec. ...	82	76	96	74	71	84	77	101	74	70	84	78	92	76	70	81	74	103	69	67	88	81	105	75	72
1961 Dec. ...	84	80	97	78	75	85	80	101	77	74	85	80	93	79	74	84	77	101	74	70	87	82	101	78	75
1962 Dec. ...	86	82	97	81	78	86	81	101	80	77	87	82	94	82	75	86	82	102	77	76	89	84	101	81	78
1963 Dec. ...	89	84	99	85	80	88	84	103	82	79	89	85	97	83	79	88	84	102	81	79	91	87	99	85	82
1964 Dec. ...	91	86	98	88	84	91	88	104	86	85	92	86	97	87	81	91	89	102	85	84	92	90	98	88	86
1965 Dec. ...	91	90	94	92	90	88	88	93	89	89	94	94	99	89	91	95	94	98	93	92	95	93	99	92	90
1966 Dec. ...	95	95	97	96	96	93	93	97	96	94	98	98	100	97	98	96	96	96	96	95	98	97	100	97	95
1967 Dec. ...	100	100	100	103	102	97	98	100	102	101	102	104	101	104	106	101	104	101	101	104	101	102	100	102	102
1968 Dec. ...	110	108	115	110	107	106	105	112	109	106	114	116	110	117	120	112	113	118	109	111	114	113	123	107	109
1969 Dec. ...	113	111	113	116	112	116	115	113	128	120	121	126	113	128	133	117	117	117	116	116	123	116	119	127	115
1970 Dec. ...	122	125	112	132	131	126	135	113	146	147	130	140	125	134	148	125	127	116	130	129	137	138	125	146	142
1971 Dec. ...	144	145	136	154	149	146	154	137	165	165	142	155	137	147	165	152	157	149	153	157	161	165	159	163	167
1972 Dec. ...	165	167	166	170	169	156	165	153	170	175	152	165	143	160	177	162	164	162	161	162	170	177	161	176	183
1973 Dec. ...	173	178	174	178	182	164	172	165	175	180	169	183	162	174	195	172	173	175	170	171	183	192	179	185	196
1974 Dec. ...	185	193	181	194	200	177	187	178	190	196	180	194	167	192	209	185	185	195	179	180	195	212	188	199	220
1975 Dec. ...	202	219	196	213	229	191	202	194	203	212	198	214	192	203	227	199	199	210	192	194	215	232	217	213	237
1976 Dec. ...	238	248	244	242	252	209	217	217	217	224	219	240	213	225	255	216	208	232	207	200	228	249	229	227	255
1977 Mar. ...	244	250	253	244	252	216	222	234	217	224	225	247	225	225	259	219	214	236	210	206	226	247	225	227	255
June ...	245	249	250	249	252	224	231	236	230	236	225	247	225	225	259	221	214	239	210	206	230	258	225	233	270
Sept. ...	252	253	267	249	252	229	234	247	230	236	234	254	235	233	264	231	222	255	217	212	234	259	229	238	270
Dec. ...	263	268	276	262	269	229	234	247	230	236	241	259	243	239	269	231	223	256	217	212	239	262	241	238	270
1978 Mar. ...	259	270	268	262	274	229	234	248	230	236	249	266	248	249	277	232	227	255	219	218	241	264	246	238	270
June ...	264	272	279	262	274	235	238	261	230	236	251	276	258	246	286	235	229	264	219	218	249	277	256	244	283
Sept. ...	286	290	308	280	288	248	251	273	245	251	252	280	260	246	291	241	236	274	222	224	250	277	257	244	283
Dec. ...	286	290	308	280	288	251	253	280	245	251	258	281	262	255	291	236	239	270	216	229	252	277	258	248	283
1979 Mar. ...	286	290	308	280	288	257	256	292	245	251	261	285	267	255	296	242	239	271	226	229	253	278	261	248	283

Reprinted with permission from the March 22, 1979 issue of Engineering News-Record magazine.

ENGINEERING NEWS-RECORD COST INDEXES IN 22 CITIES

	PITTSBURGH					ST. LOUIS					SAN FRANCISCO				
	BCI	CCI	MCC	SLI	CLI	BCI	CCI	MCC	SLI	CLI	BCI	CCI	MCC	SLI	CLI
1954 Dec.	65	59	83	57	52	70	57	89	56	47	59	51	74	49	44
1955 Dec.	69	62	91	59	55	74	60	96	58	49	62	54	77	52	47
1956 Dec.	72	65	98	61	56	77	63	100	61	51	65	57	80	55	50
1957 Dec.	77	71	102	66	62	80	66	105	63	54	68	60	81	60	54
1958 Dec.	81	74	105	70	66	82	70	105	66	59	72	64	86	62	57
1959 Dec.	82	76	105	72	69	84	72	105	69	62	74	67	86	65	61
1960 Dec.	85	80	107	77	73	84	75	104	71	65	74	69	83	68	65
1961 Dec.	86	82	107	79	77	85	77	102	74	70	76	72	83	71	68
1962 Dec.	88	85	107	82	81	88	81	102	78	74	81	75	85	78	71
1963 Dec.	91	89	109	86	85	91	85	105	81	79	85	80	90	81	77
1964 Dec.	93	92	109	91	90	93	88	106	85	83	86	83	88	85	81
1965 Dec.	91	90	95	97	93	95	91	105	88	87	92	90	92	91	89
1966 Dec.	92	94	97	98	98	97	96	98	96	95	96	96	96	97	95
1967 Dec.	98	96	97	109	100	101	102	102	101	102	103	104	102	103	104
1968 Dec.	113	110	118	120	112	111	112	117	107	110	112	112	116	109	110
1969 Dec.	114	124	117	123	133	117	123	115	118	126	112	117	107	116	119
1970 Dec.	125	128	126	136	135	124	137	116	131	143	125	126	110	134	131
1971 Dec.	149	149	153	160	155	140	154	136	143	159	145	146	143	147	147
1972 Dec.	163	165	177	168	169	154	166	153	155	170	160	160	156	163	160
1973 Dec.	167	174	179	174	181	161	176	162	161	181	168	177	168	168	179
1974 Dec.	178	185	193	184	191	173	194	173	174	200	187	198	185	188	202
1975 Dec.	190	201	210	194	208	190	211	192	189	217	203	217	192	210	224
1976 Dec.	208	223	237	207	229	207	228	215	202	233	227	245	226	228	250
1977 Mar.	210	224	238	210	229	210	230	222	202	233	230	247	233	228	250
June	210	223	237	210	229	214	231	226	206	233	231	247	233	230	250
Sept.	224	240	251	225	247	227	242	247	213	240	231	247	233	230	250
Dec.	226	241	251	225	247	225	244	244	213	244	238	249	242	235	250
1978 Mar.	228	243	263	225	247	230	246	252	214	244	240	262	245	237	267
June	232	245	271	225	247	233	253	258	216	251	248	270	265	237	270
Sept.	245	264	288	236	268	238	254	264	220	251	261	270	265	259	270
Dec.	246	265	291	236	268	243	257	264	228	255	261	270	265	259	270
1979 Mar.	246	265	292	236	268	246	259	271	228	255	274	284	298	259	279

	SEATTLE					CANADA — MONTREAL					CANADA — TORONTO				
	BCI	CCI	MCC	SLI	CLI	BCI	CCI	MCC	SLI	CLI	BCI	CCI	MCC	SLI	CLI
1954 Dec.	65	60	80	56	54	71	57	92	47	39	—	—	—	—	—
1955 Dec.	68	61	83	57	54	72	58	92	50	41	71	60	83	57	47
1956 Dec.	71	64	88	59	57	74	59	96	50	41	75	66	89	60	53
1957 Dec.	73	66	87	63	60	77	62	97	54	45	77	67	90	63	54
1958 Dec.	77	71	92	66	64	76	64	94	55	48	79	70	90	66	60
1959 Dec.	79	74	93	69	68	78	66	93	60	53	80	72	89	70	63
1960 Dec.	80	76	90	73	71	79	68	92	63	57	81	76	86	75	70
1961 Dec.	82	79	92	76	75	78	70	90	65	60	82	77	87	77	72
1962 Dec.	85	81	92	80	78	80	73	90	69	64	83	80	88	79	75
1963 Dec.	88	85	95	83	82	82	76	90	73	69	84	82	87	82	79
1964 Dec.	90	87	95	86	85	84	78	90	77	73	87	86	90	85	84
1965 Dec.	94	91	97	91	90	84	81	87	81	78	88	89	84	92	91
1966 Dec.	97	96	97	97	95	97	96	100	93	94	98	98	99	97	97
1967 Dec.	102	100	101	103	100	103	106	100	107	110	104	107	102	105	110
1968 Dec.	113	111	115	112	110	107	110	101	114	115	111	115	108	115	118
1969 Dec.	119	118	112	123	120	125	125	124	126	126	117	124	104	131	134
1970 Dec.	127	125	116	135	127	124	129	114	135	137	124	137	98	152	158
1971 Dec.	142	139	135	146	140	129	136	120	140	144	138	158	105	173	186
1972 Dec.	152	149	147	156	149	137	146	127	149	155	163	185	135	194	212
1973 Dec.	165	163	168	164	162	158	166	153	167	172	172	203	136	211	238
1974 Dec.	184	186	184	184	186	186	195	188	185	198	180	218	137	226	261
1975 Dec.	206	214	211	203	215	199	221	183	218	240	206	244	161	253	292
1976 Dec.	227	236	239	219	235	219	245	202	239	266	266	277	167	290	335
1977 Mar.	230	238	247	219	235	227	258	206	251	284	233	283	181	290	335
June	228	237	243	219	235	240	270	206	279	301	245	296	185	310	354
Sept.	246	254	256	238	253	242	271	209	279	301	251	300	197	310	354
Dec.	245	254	256	238	253	242	271	209	279	301	251	311	190	316	375
1978 Mar.	251	255	262	244	253	249	275	223	279	301	258	316	203	316	375
June	269	261	289	256	253	261	275	223	304	301	261	317	207	321	375
Sept.	279	283	301	264	277	282	288	262	304	301	272	329	223	326	385
Dec.	279	283	302	264	277	282	288	262	304	301	280	342	229	336	400
1979 Mar.	280	284	304	264	277	286	291	270	304	301	283	344	235	336	400

Reprinted with permission from the March 22, 1979 issue of Engineering News-Record magazine.

APPENDIX 4

RELATIVE ROUGHNESS OF PIPE MATERIALS AND FRICTION FACTORS FOR TURBULENT FLOW

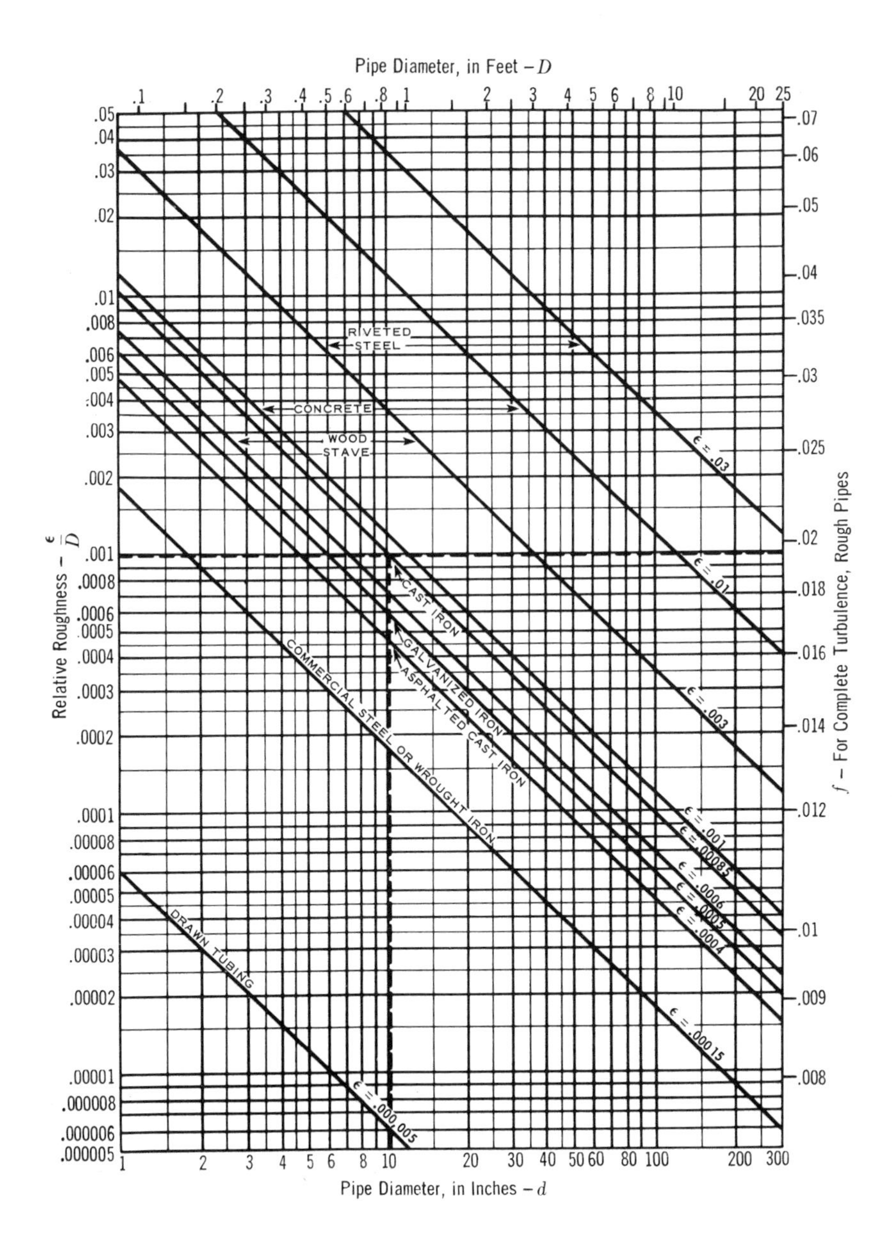

Reprinted with permission from Technical Paper No. 410, Crane Company, New York, N.Y.

FRICTION FACTORS FOR ANY TYPE COMMERCIAL PIPE

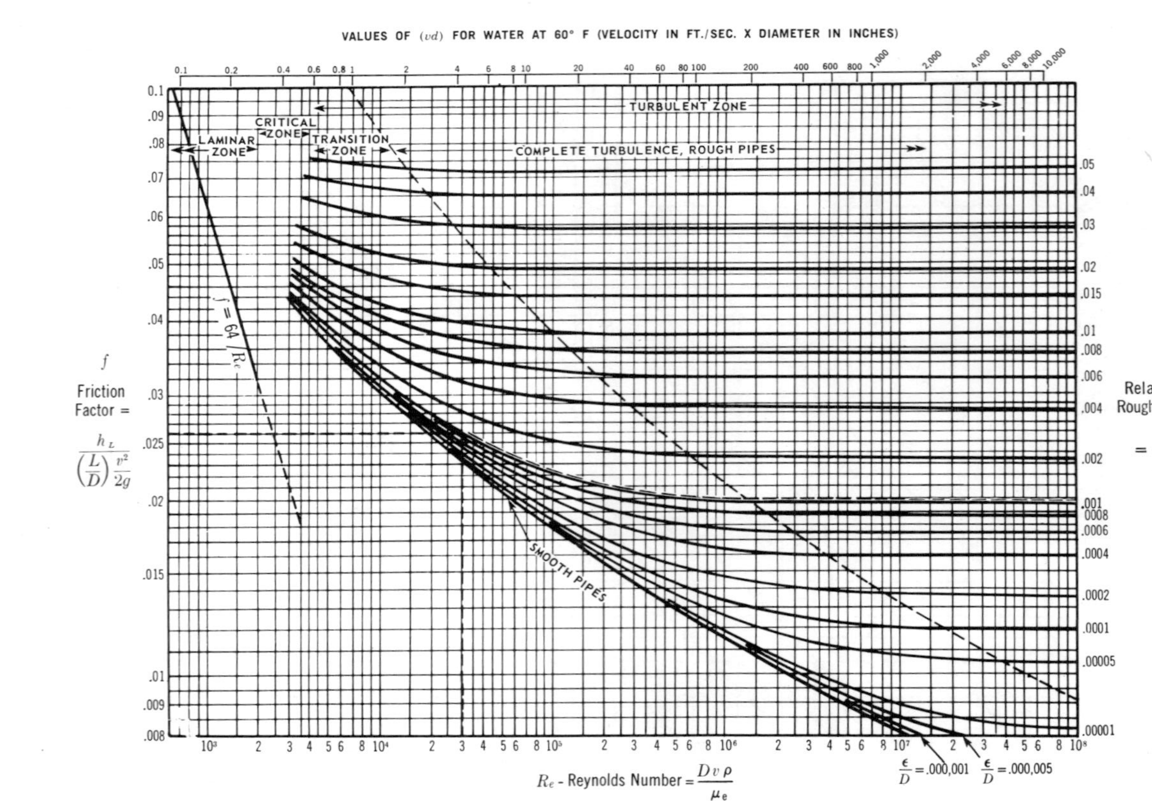

Reprinted with permission from Technical Paper No. 410, Crane Company, New York, N.Y.

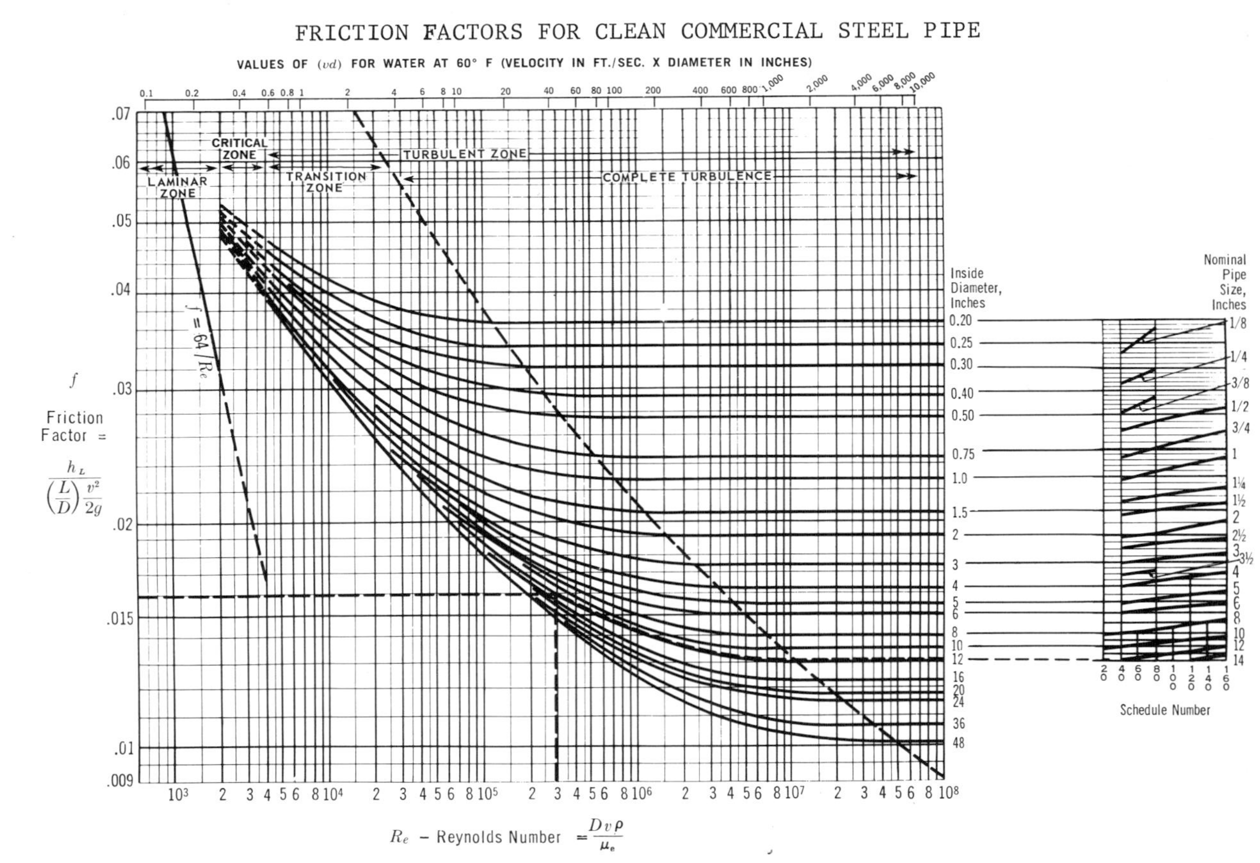

Reprinted with permission from Technical Paper No. 410, Crane Company, New York, N.Y.

APPENDIX 5

HAZEN-WILLIAMS FORMULA PIPE FLOW CHART

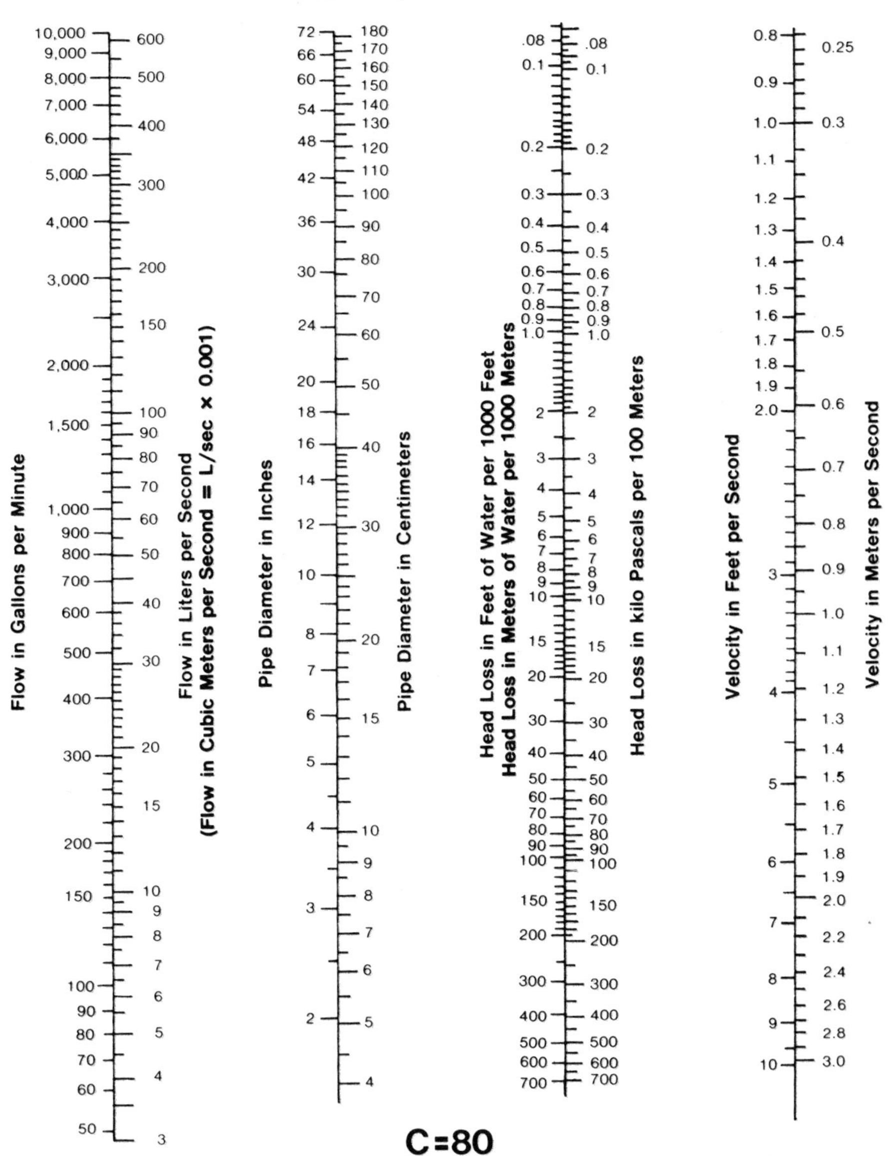

Reprinted with permission from the April 1977 issue of Water & Sewage Works magazine. Prepared by Frank Reid and Harold Stone, Water & Sewage works staff.

HAZEN-WILLIAMS FORMULA PIPE FLOW CHART

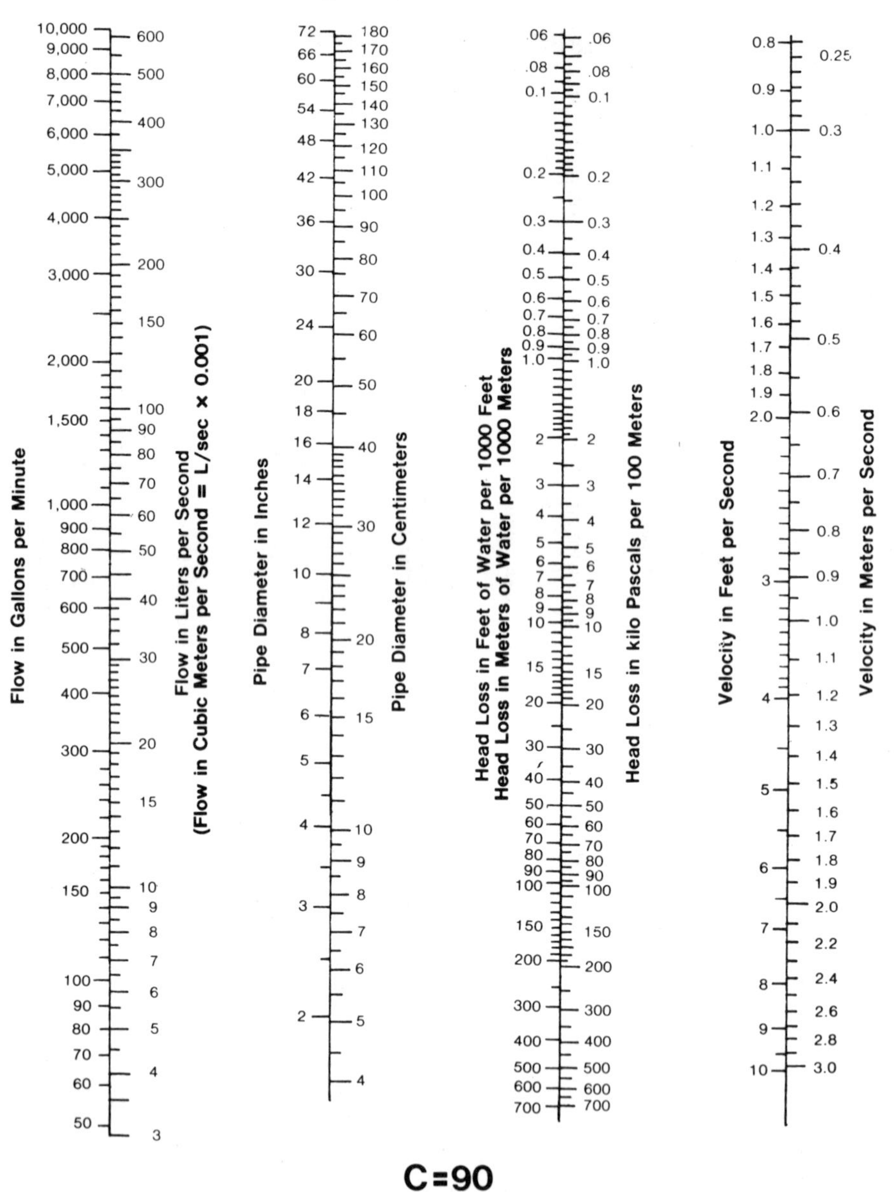

Reprinted with permission from the April 1977 issue of Water & Sewage Works magazine. Prepared by Frank Reid and Harold Stone, Water & Sewage Works staff.

HAZEN-WILLIAMS FORMULA PIPE FLOW CHART

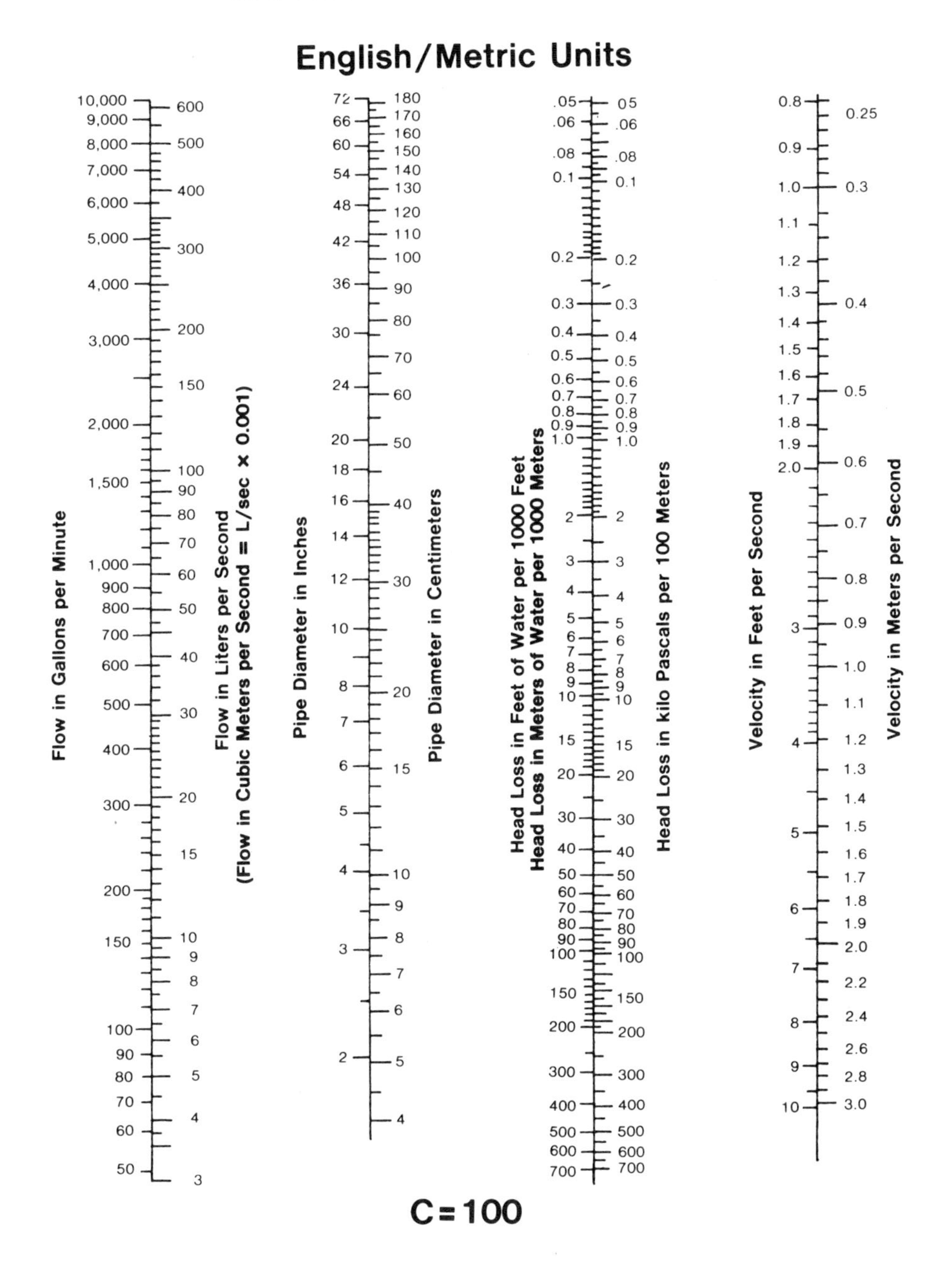

Reprinted with permission from the April 1977 issue of Water & Sewage Works magazine. Prepared by Frank Reid and Harold Stone, Water & Sewage Works staff.

HAZEN-WILLIAMS FORMULA PIPE FLOW CHART

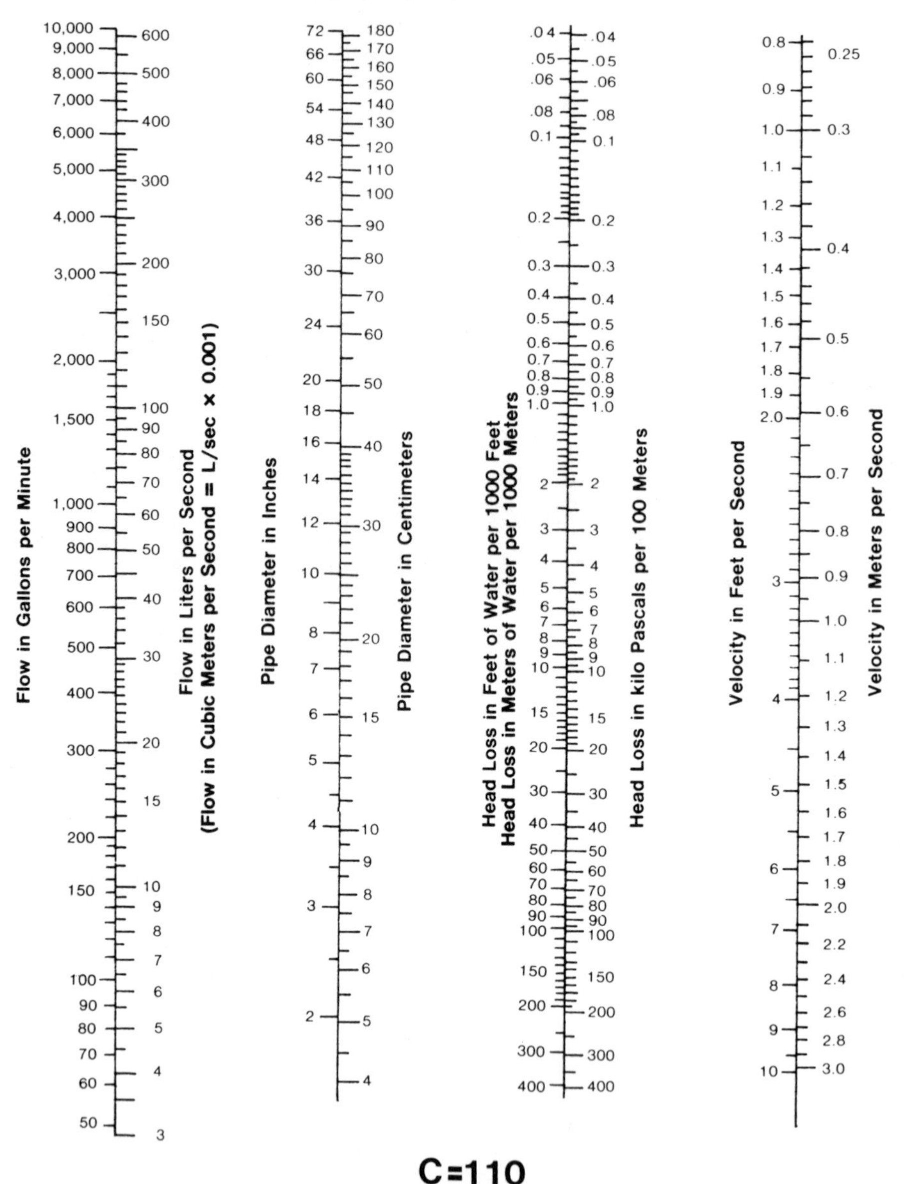

Reprinted with permission from the April 1977 issue of Water & Sewage Works magazine. Prepared by Frank Reid and Harold Stone, Water & Sewage Works staff.

HAZEN-WILLIAMS FORMULA PIPE FLOW CHART

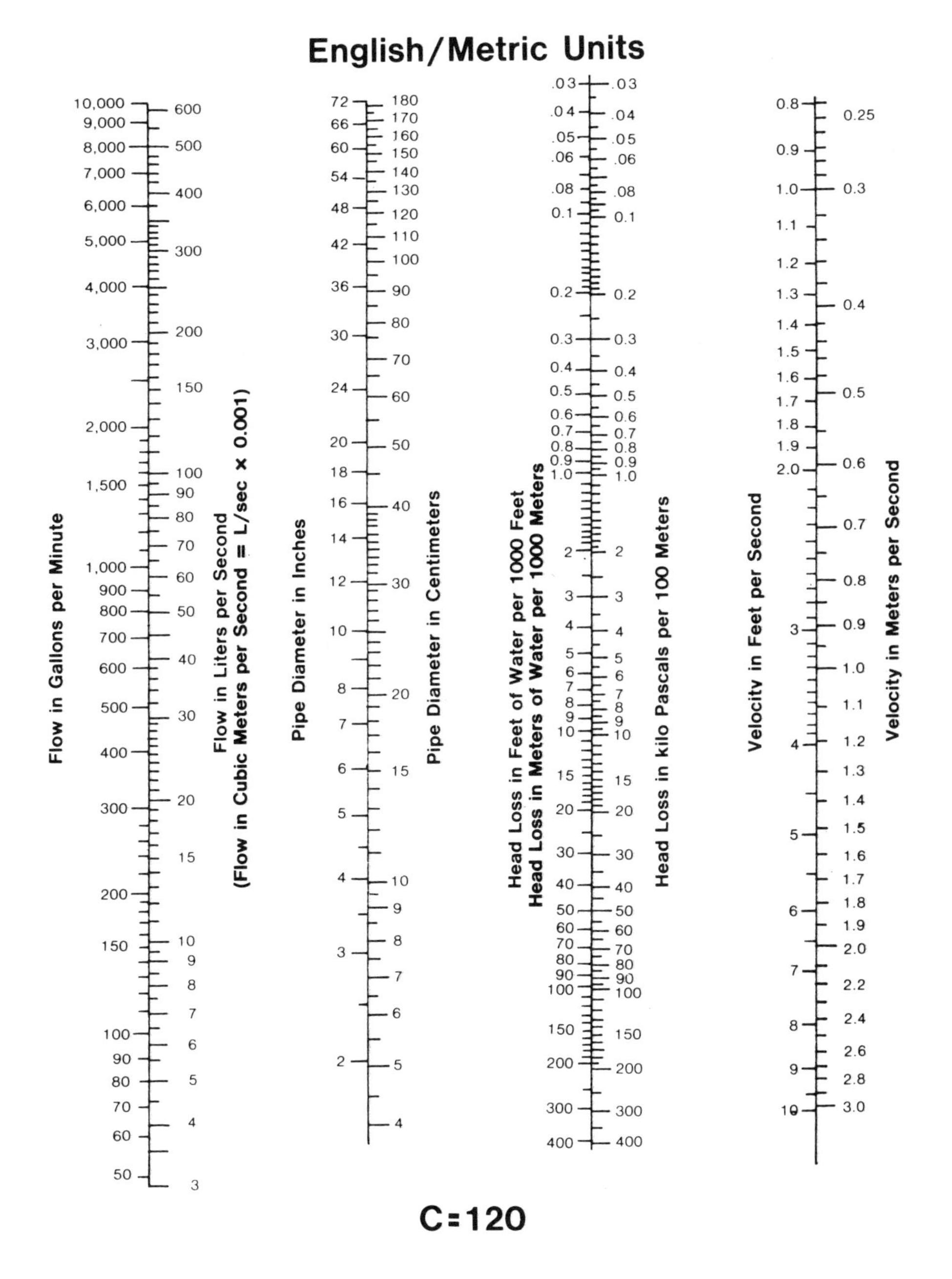

Reprinted with permission from the April 1977 issue of Water & Sewage Works magazine. Prepared by Frank Reid and Harold Stone, Water & Sewage Works staff.

HAZEN-WILLIAMS FORMULA PIPE FLOW CHART

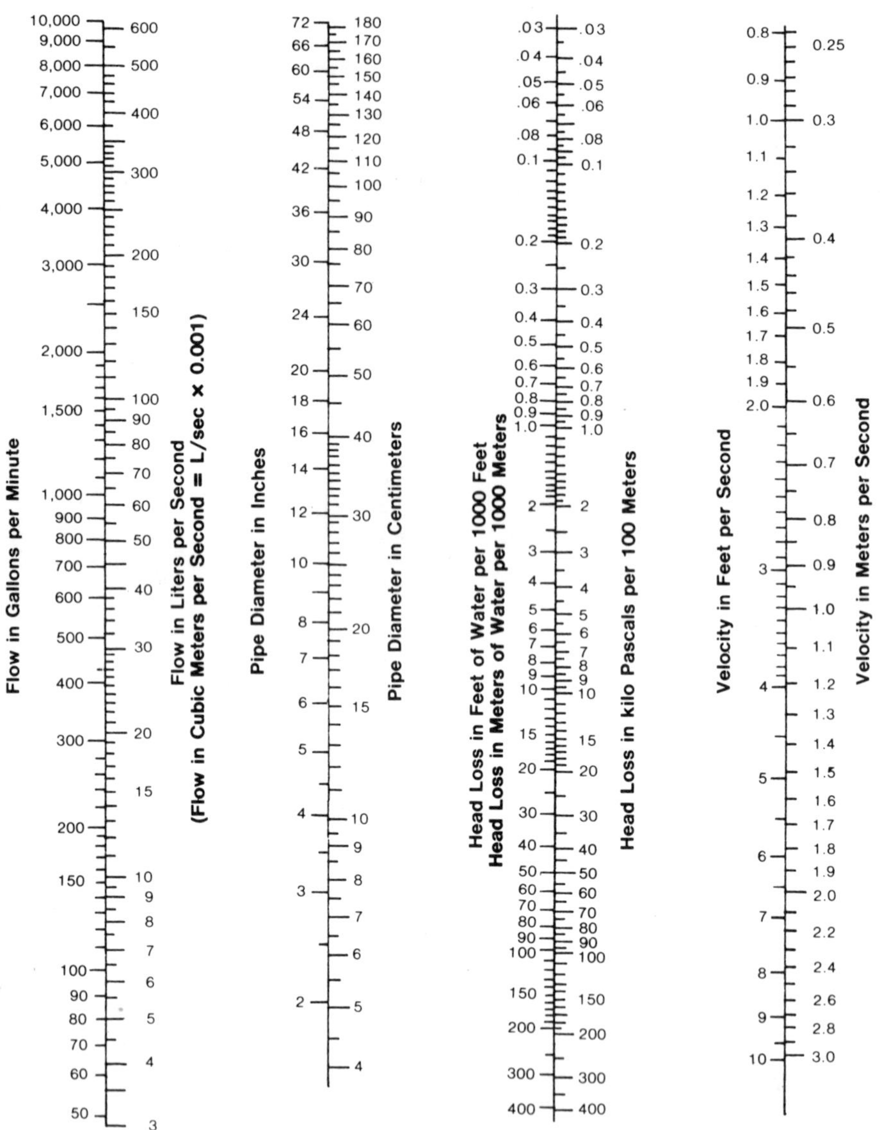

C=130

Reprinted with permission from the April 1977 issue of Water & Sewage Works magazine. Prepared by Frank Reid and Harold Stone, Water & Sewage Works staff.

HAZEN-WILLIAMS FORMULA PIPE FLOW CHART

English/Metric Units

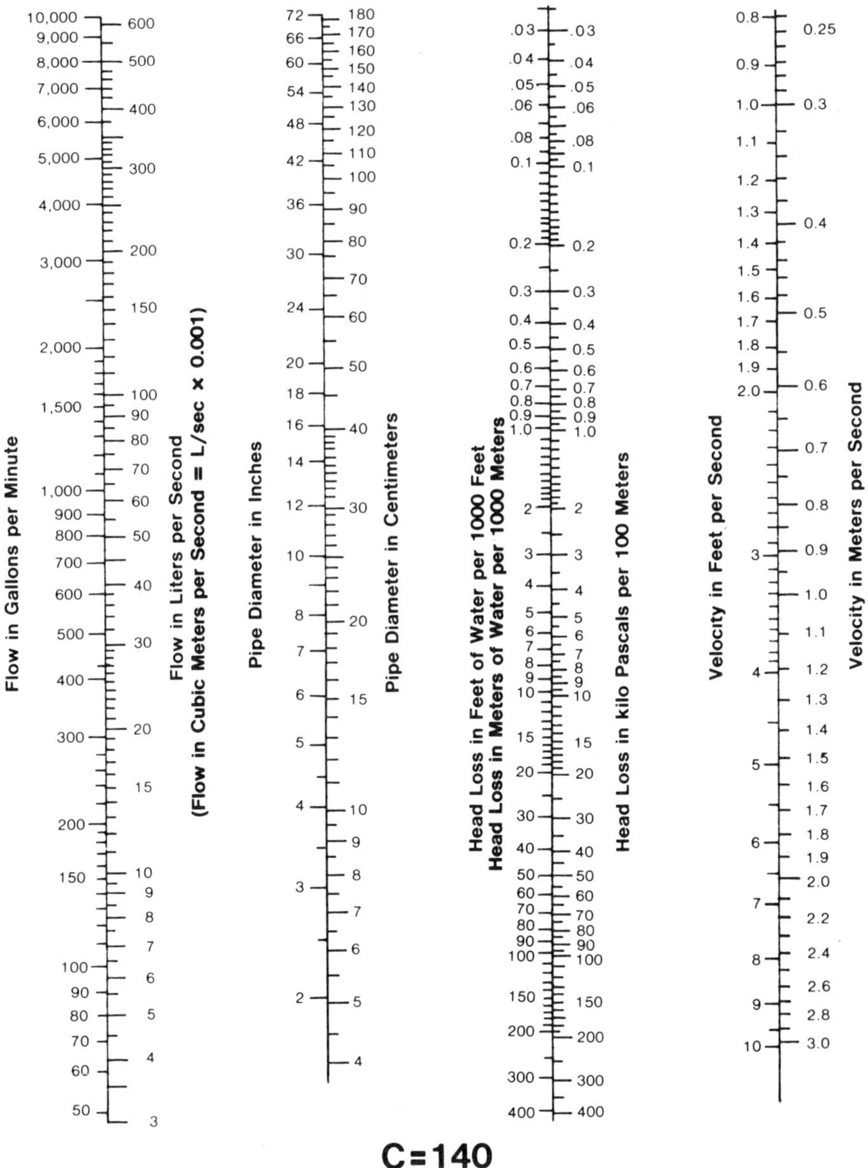

Reprinted with permission from the April 1977 issue of Water & Sewage Works magazine. Prepared by Frank Reid and Harold Stone, Water & Sewage Works staff.

HAZEN-WILLIAMS FORMULA PIPE FLOW CHART

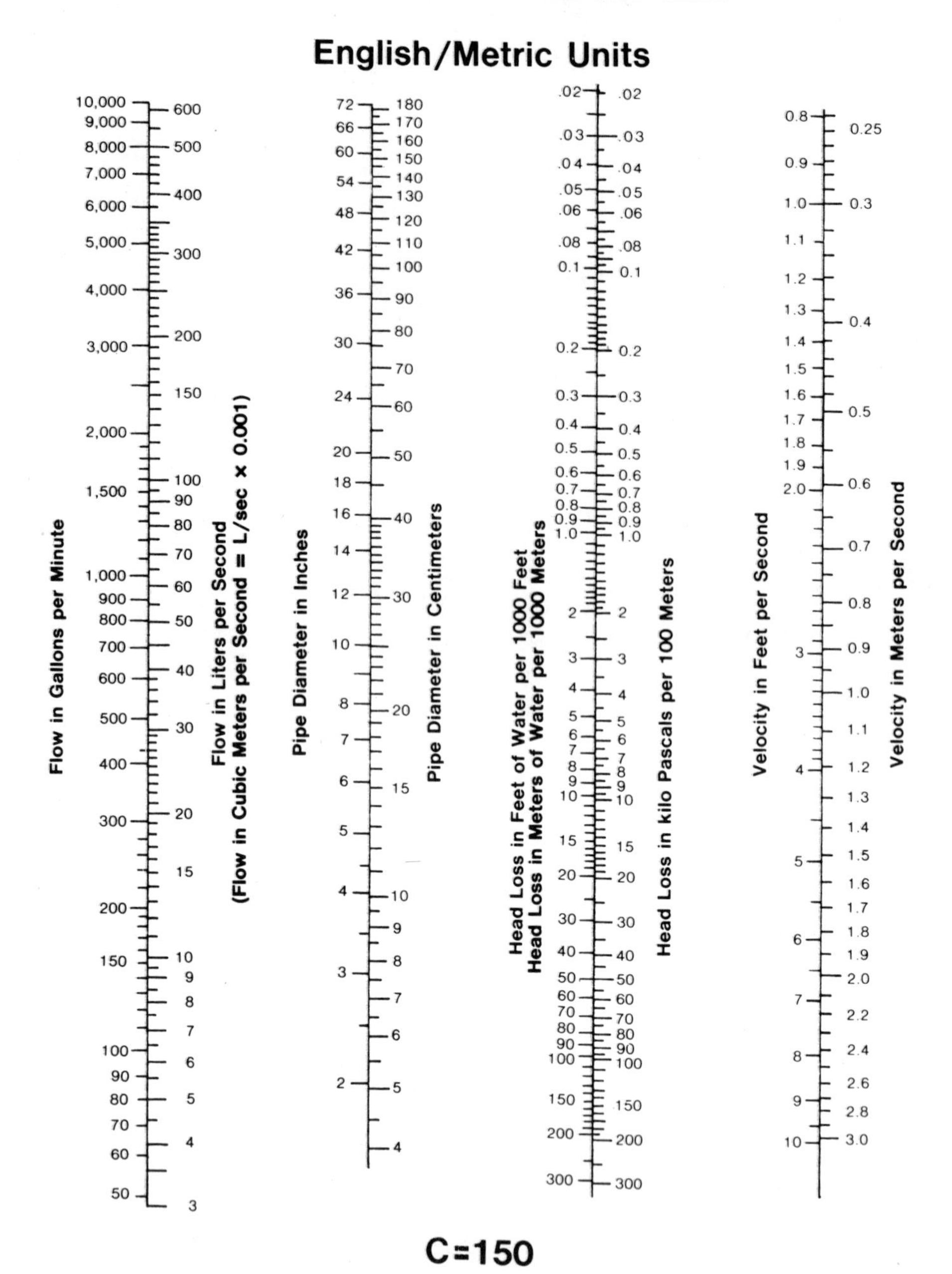

Reprinted with permission from the April 1977 issue of Water & Sewage Works magazine. Prepared by Frank Reid and Harold Stone, Water & Sewage Works staff.

APPENDIX 6

MANNING FORMULA PIPE FLOW CHART

English/Metric Units

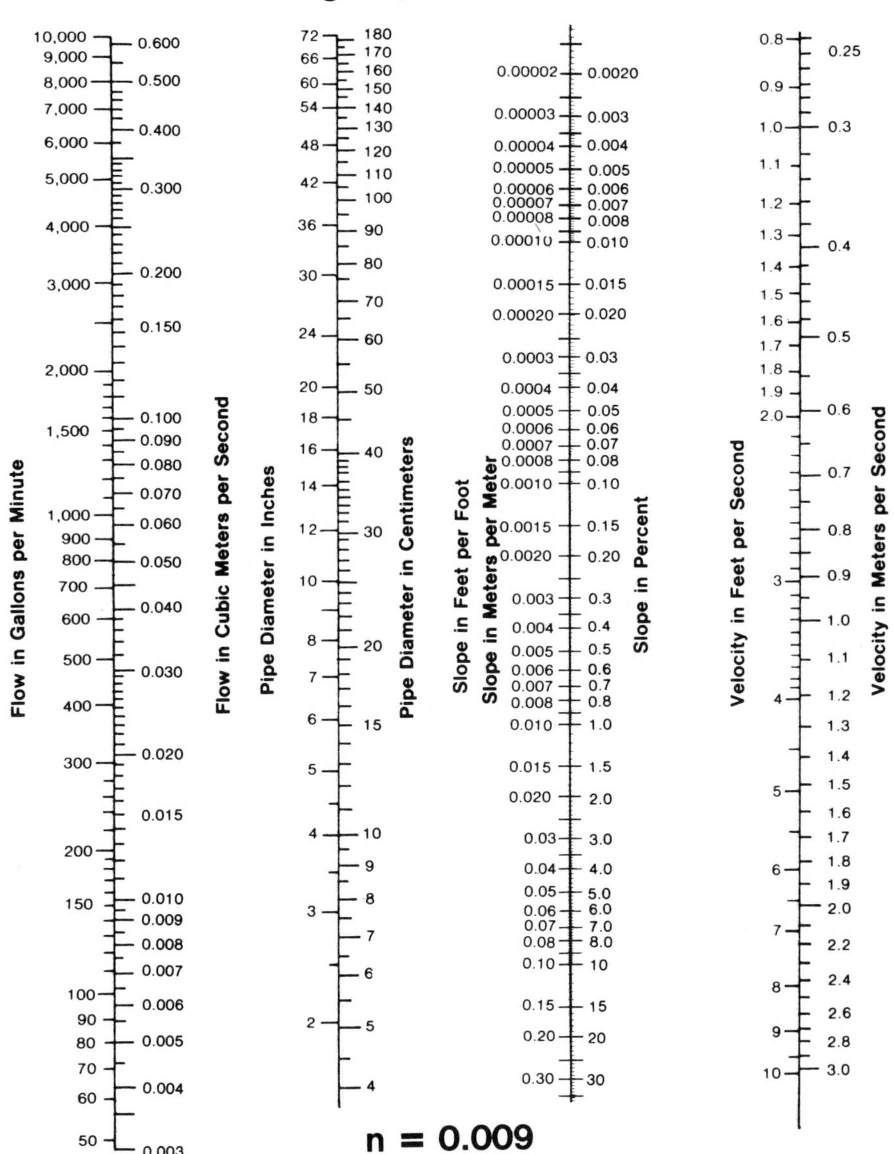

Chart based on the formula $Q = \frac{1.486}{n} \times AR^{\frac{2}{3}} \times S^{\frac{1}{2}}$ for pipe flowing full.

Reprinted with permission from the April 30, 1978 reference issue of Water & Sewage Works magazine. Prepared by Frank Reid and Harold Stone, Water & Sewage Works staff.

MANNING FORMULA PIPE FLOW CHART

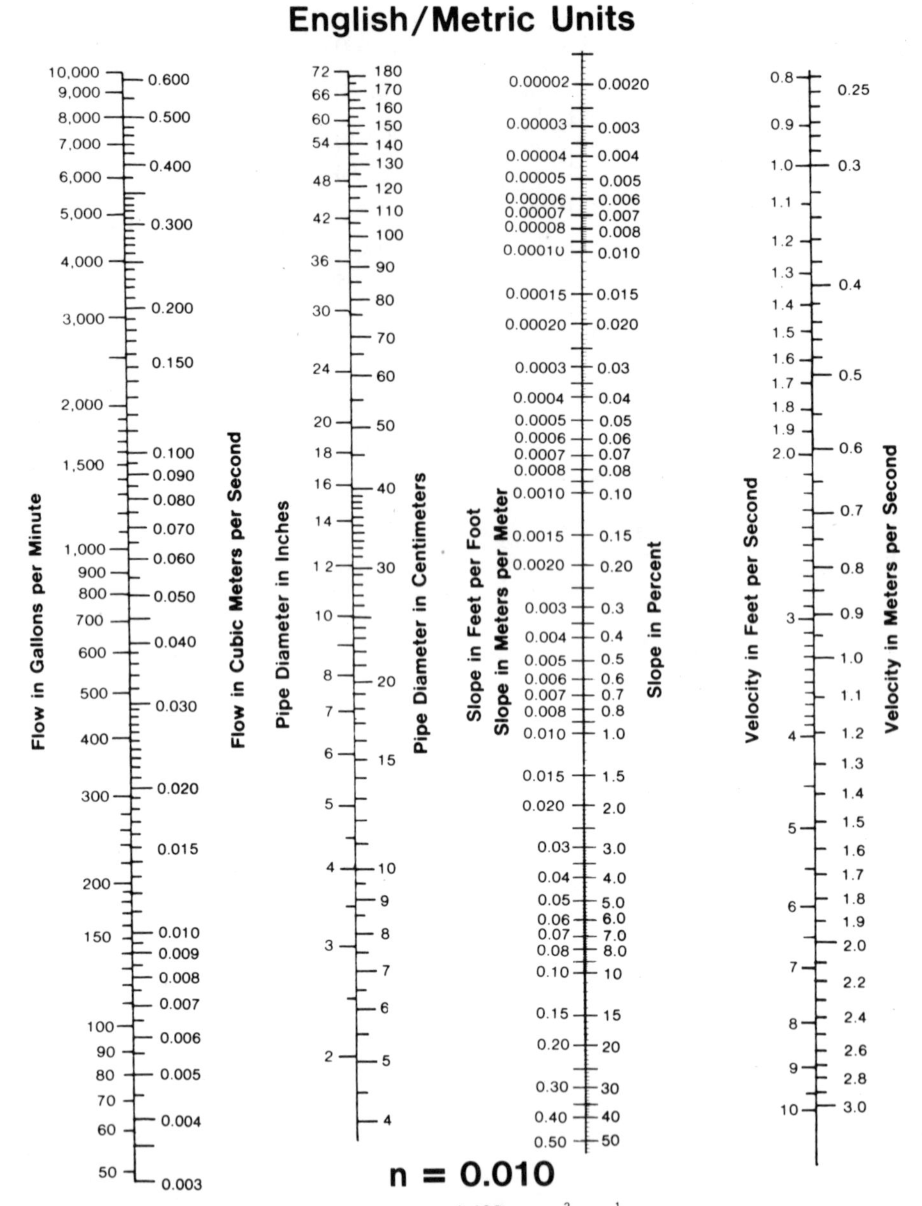

Chart based on the formula $Q = \frac{1.486}{n} \times AR^{\frac{2}{3}} \times S^{\frac{1}{2}}$ for pipe flowing full.

Reprinted with permission from the April 30, 1978 reference issue of Water & Sewage Works magazine. Prepared by Frank Reid and Harold Stone, Water & Sewage Works staff.

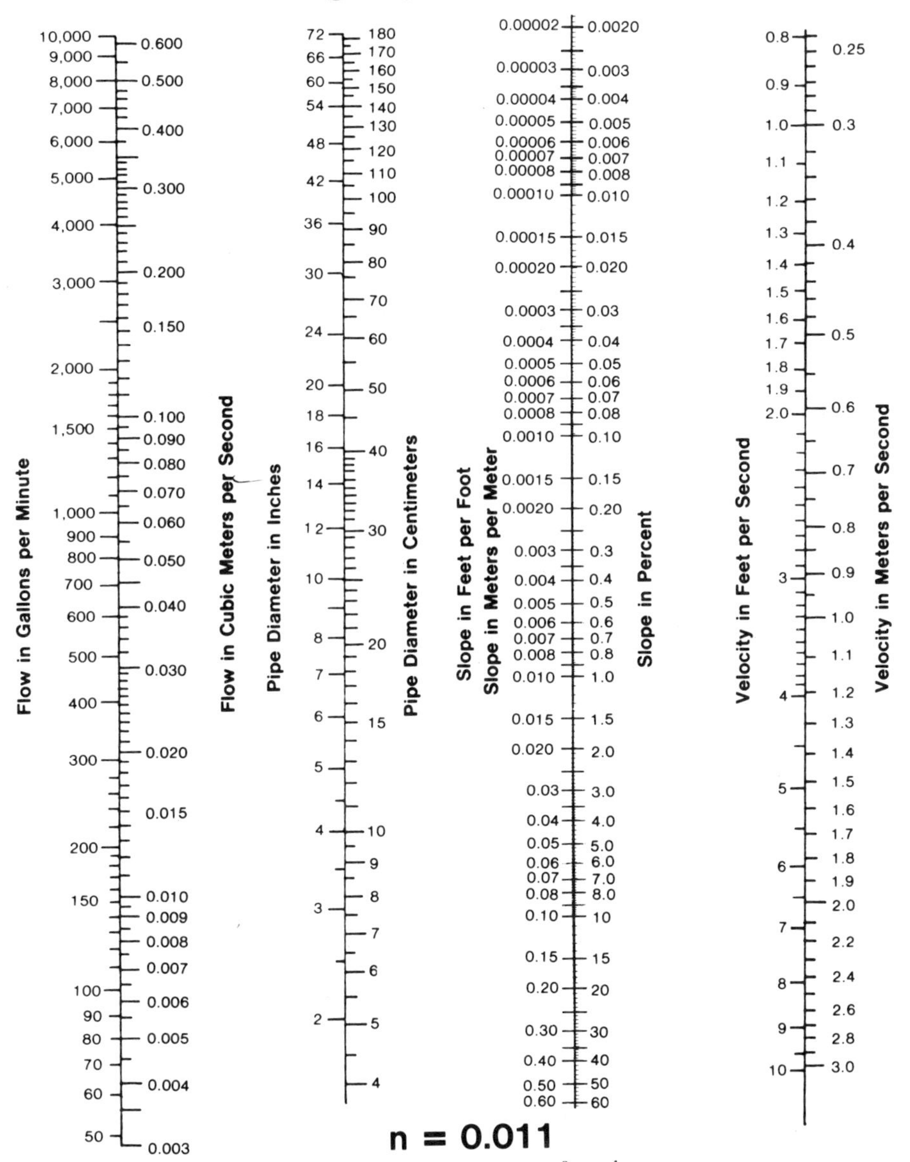

Chart based on the formula $Q = \frac{1.486}{n} \times AR^{\frac{2}{3}} \times S^{\frac{1}{2}}$ for pipe flowing full.

Reprinted with permission from the April 30, 1978 reference issue of Water & Sewage Works magazine. Prepared by Frank Reid and Harold Stone, Water & Sewage Works staff.

English/Metric Units

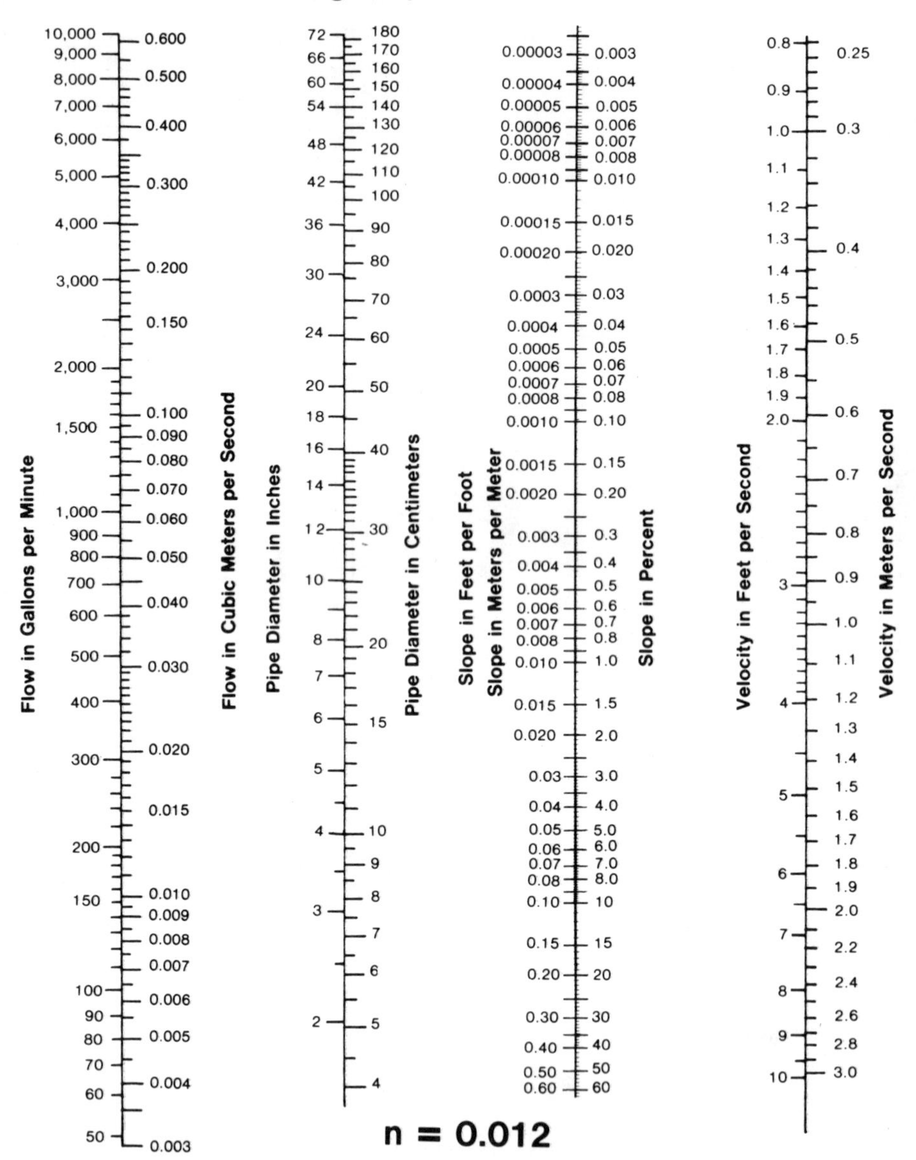

Chart based on the formula $Q = \frac{1.486}{n} \times AR^{\frac{2}{3}} \times S^{\frac{1}{2}}$ for pipe flowing full.

Reprinted with permission from the April 30, 1978 reference issue of Water & Sewage Works magazine. Prepared by Frank Reid and Harold Stone, Water & Sewage Works staff.

MANNING FORMULA PIPE FLOW CHART

English/Metric Units

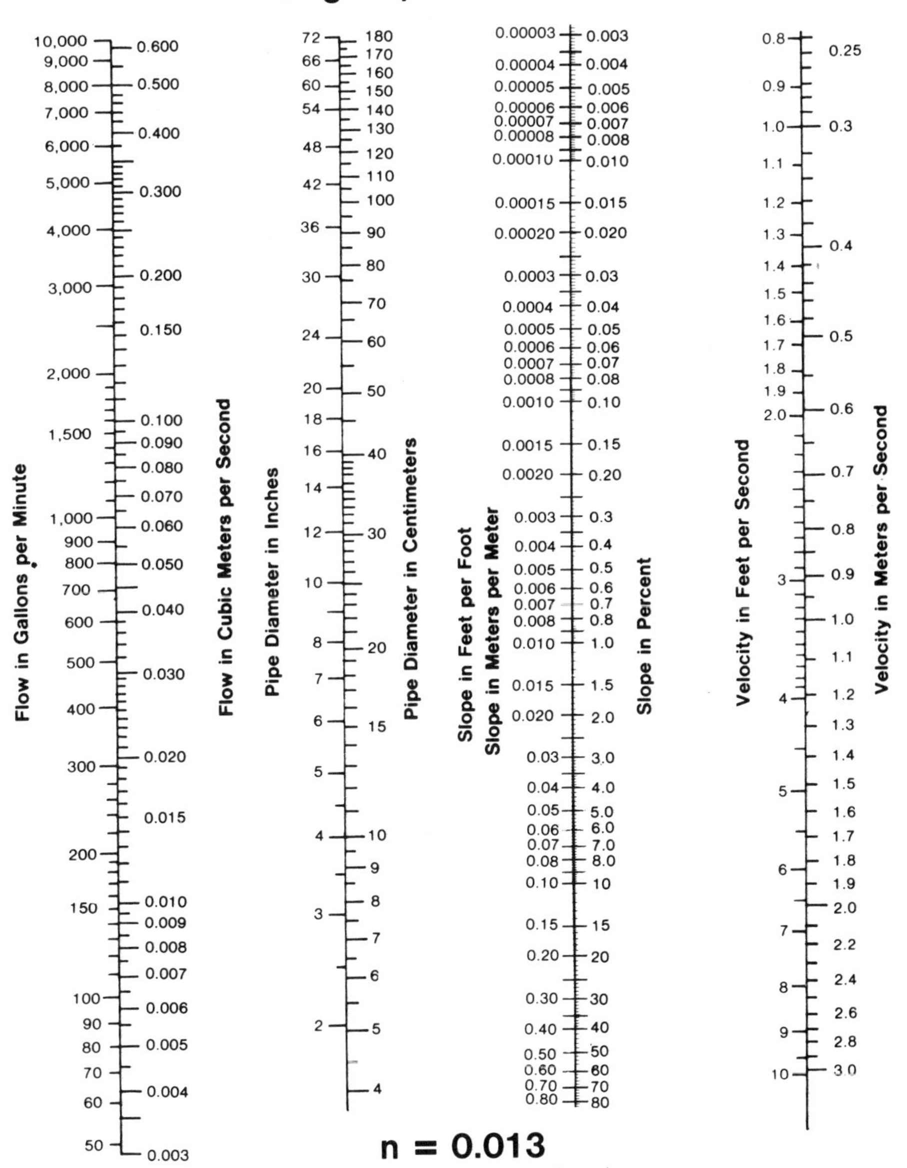

Chart based on the formula $Q = \frac{1.486}{n} \times AR^{\frac{2}{3}} \times S^{\frac{1}{2}}$ for pipe flowing full.

Reprinted with permission from the April 30, 1978 reference issue of Water & Sewage Works magazine. Prepared by Frank Reid and Harold Stone, Water & Sewage Works staff.

MANNING FORMULA PIPE FLOW CHART

English/Metric Units

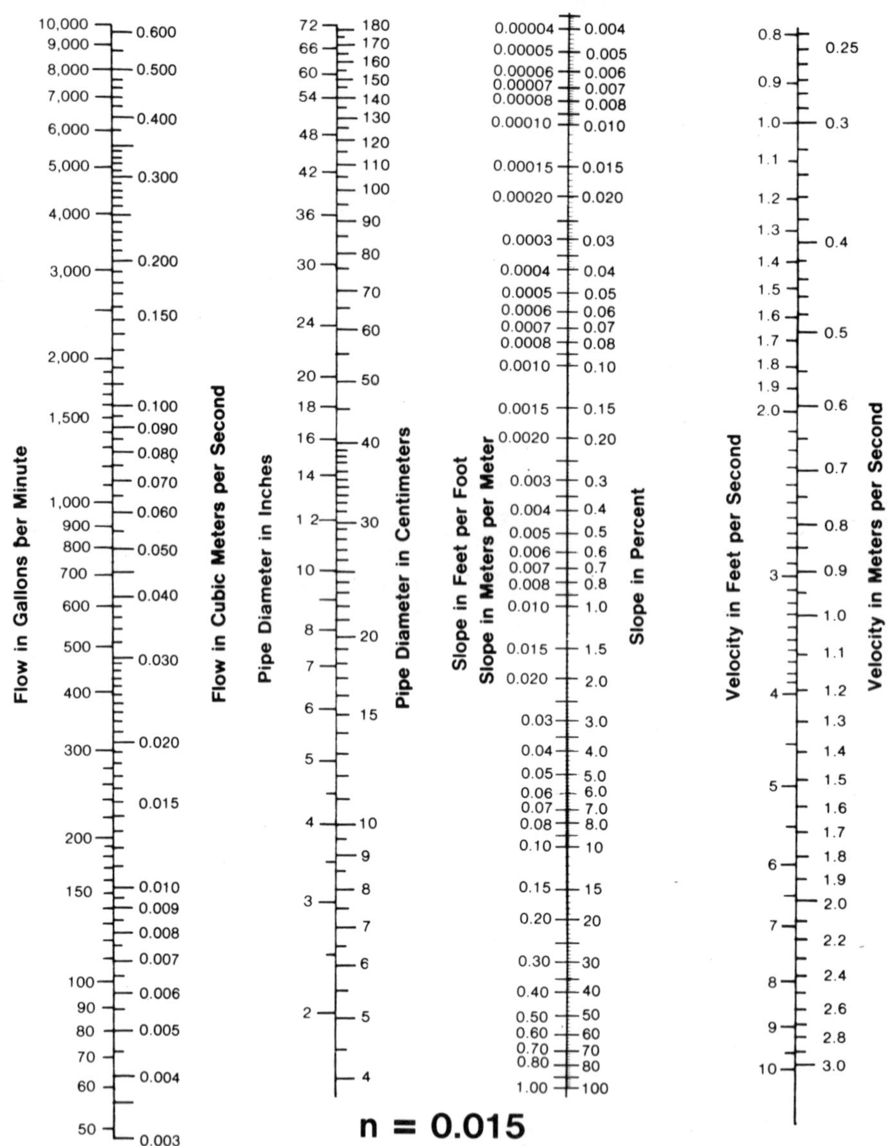

Chart based on the formula $Q = \frac{1.486}{n} \times AR^{\frac{2}{3}} \times S^{\frac{1}{2}}$ for pipe flowing full.

Reprinted with permission from the April 30, 1978 reference issue of Water & Sewage Works magazine. Prepared by Frank Reid and Harold Stone, Water & Sewage Works staff.

MANNING FORMULA PIPE FLOW CHART

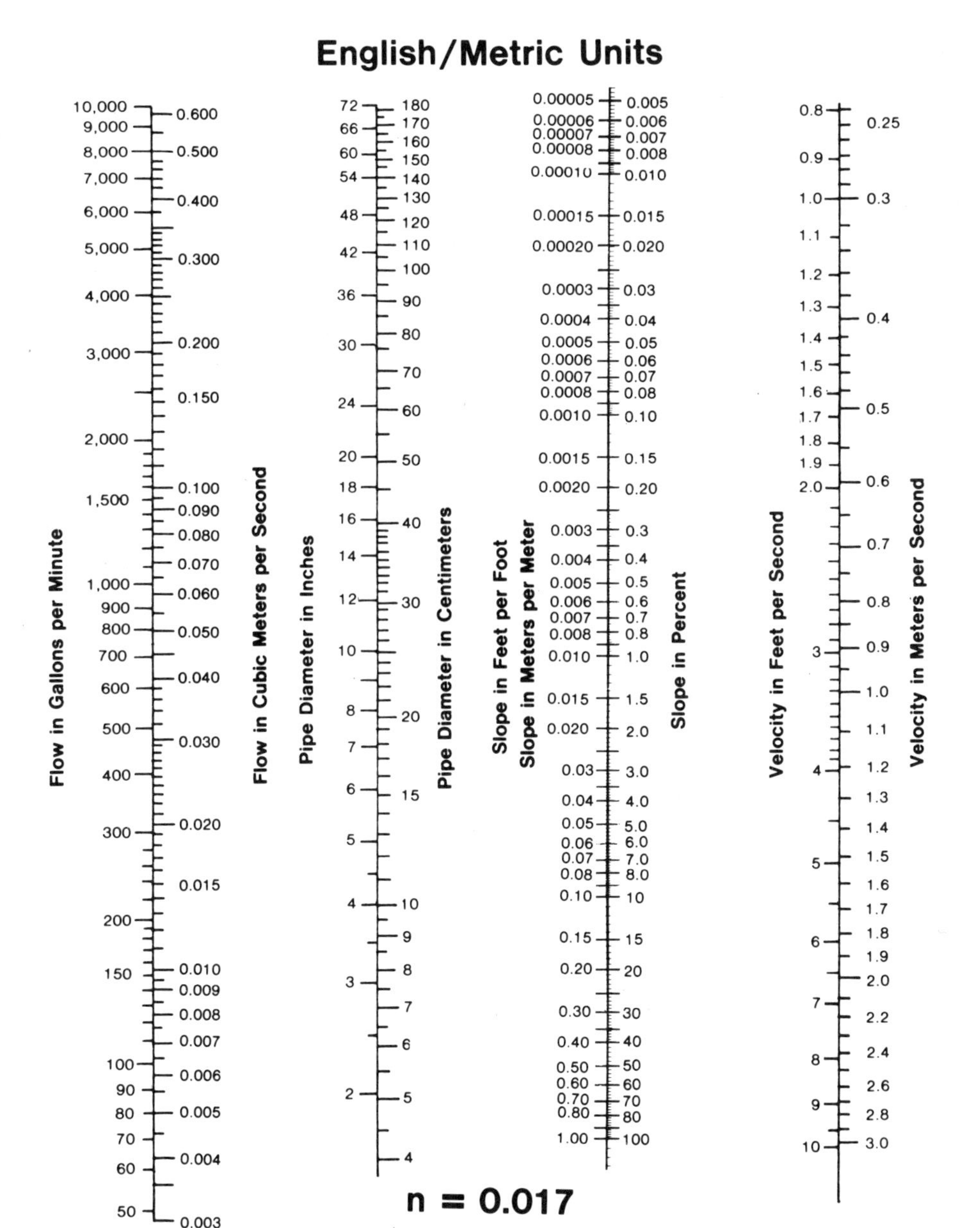

Chart based on the formula $Q = \frac{1.486}{n} \times AR^{\frac{2}{3}} \times S^{\frac{1}{2}}$ for pipe flowing full.

Reprinted with permission from the April 30, 1978 reference issue of Water & Sewage Works magazine. Prepared by Frank Reid and Harold Stone, Water & Sewage Works staff.

MANNING FORMULA PIPE FLOW CHART

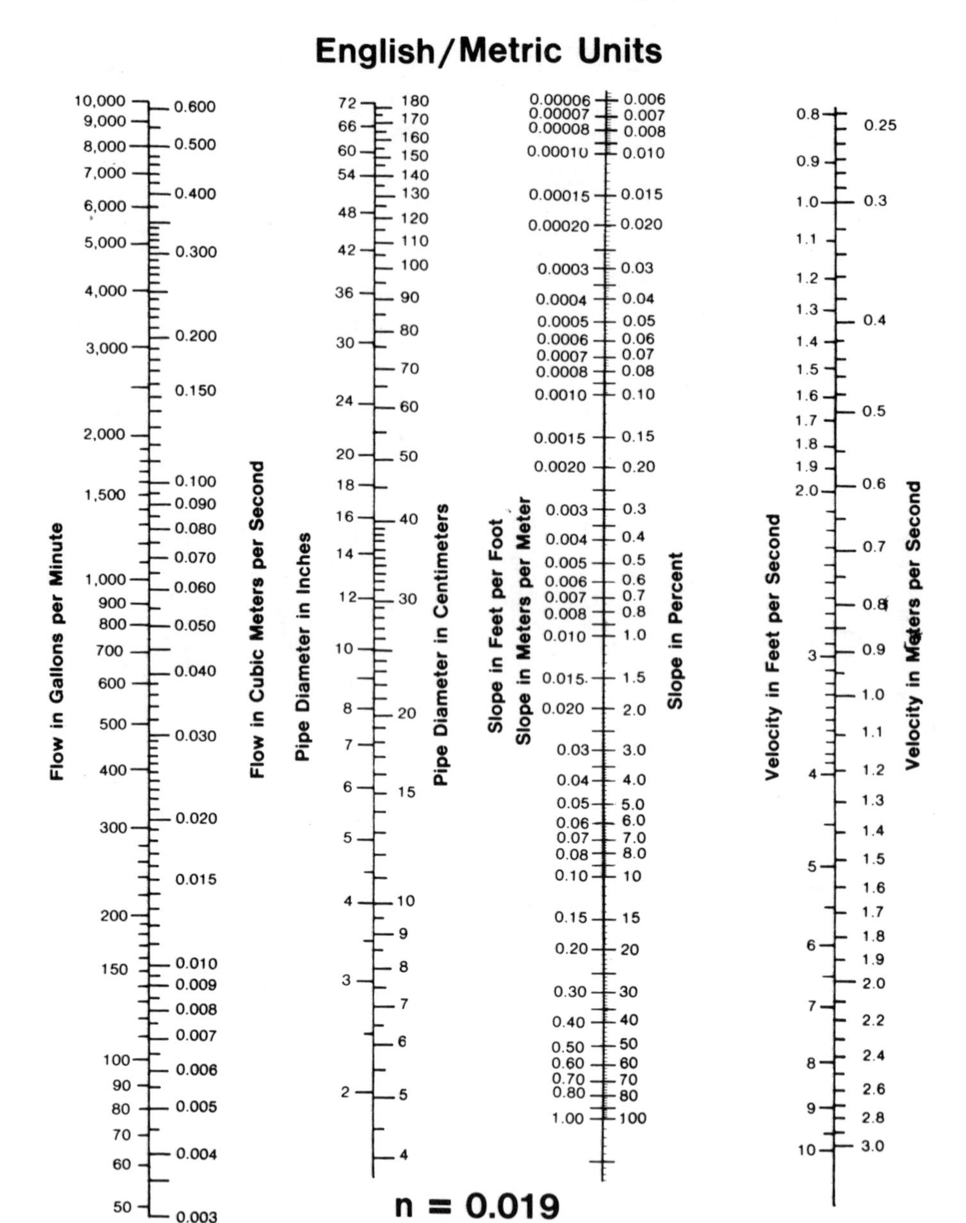

Chart based on the formula $Q = \frac{1.486}{n} \times AR^{\frac{2}{3}} \times S^{\frac{1}{2}}$ for pipe flowing full.

Reprinted with permission from the April 30, 1978 reference issue of Water & Sewage Works magazine. Prepared by Frank Reid and Harold Stone, Water & Sewage Works staff.

INDEX